CAUSATION, CHANCE, AND CREDENCE

THE UNIVERSITY OF WESTERN ONTARIO SERIES IN PHILOSOPHY OF SCIENCE

A SERIES OF BOOKS
IN PHILOSOPHY OF SCIENCE, METHODOLOGY,
EPISTEMOLOGY, LOGIC, HISTORY OF SCIENCE,
AND RELATED FIELDS

VOLUME 41

CAUSATION, CHANCE, AND CREDENCE

Proceedings of the Irvine Conference on Probability and Causation

VOLUME 1

Edited by

BRIAN SKYRMS

Department of Philosophy, University of California at Irvine

and

WILLIAM L. HARPER

Talbot College, The University of Western Ontario

KLUWER ACADEMIC PUBLISHERS

DORDRECHT / BOSTON / LONDON

Library of Congress Cataloging-in-Publication Data

CIP

Irvine Conference on Probability and Causation (1985: University of California at Irvine)
Causation, chance, and credence.

(The University of Western Ontario series in philosophy of science; v. 41)
Includes index.
1. Causality (Physics) — Congresses. 2. Probabilities — Congresses.
3. Science — Philosophy — Congresses. I. Skyrms, Brian. II. Harper, William L. (William Leonard), 1943– . III. Title. IV. Series.
QC6.4.C3I78 1985 530′.01 87–28857
ISBN 90–277–2633–7 (v. 1)

Published by Kluwer Academic Publishers,
P.O. Box 17, 3300 AA Dordrecht, The Netherlands.

Kluwer Academic Publishers incorporates the publishing programmes of
D. Reidel, Martinus Nijhoff, Dr W. Junk and MTP Press.

Sold and Distributed in the U.S.A. and Canada
by Kluwer Academic Publishers,
101 Philip Drive, Norwell, MA 02061, U.S.A.

In all other countries, sold and distributed
by Kluwer Academic Publishers Group,
P.O. Box 322, 3300 AH Dordrecht, The Netherlands.

Printed in The Netherlands

TABLE OF CONTENTS

CONTENTS OF VOLUME II

PREFACE

The papers collected here are, with three exceptions, those presented at a conference on probability and causation held at the University of California at Irvine on July 15–19, 1985. The exceptions are that David Freedman and Abner Shimony were not able to contribute the papers that they presented to this volume, and that Clark Glymour who was not able to attend the conference did contribute a paper. We would like to thank the National Science Foundation and the School of Humanities of the University of California at Irvine for generous support.

WILLIAM HARPER — *University of Western Ontario*

BRIAN SKYRMS — *University of California at Irvine*

INTRODUCTION TO CAUSATION, CHANCE, AND CREDENCE

The search for causes is so central to science that it has sometimes been taken as the defining attribute of the scientific enterprise. Yet even after twenty-five centuries of philosophical analysis the meaning of "cause" is still a matter of controversy, among scientists as well as philosophers.

Part of the problem is that the servicable concepts of causation built out of Necessity, Sufficiency, Locality, and Temporal Precedence were constructed for a deterministic world-view which has been obsolete since the advent of quantum theory. A physically credible theory of causation must be, at basis, statistical. And statistical analyses of causation may be of interest even when an underlying deterministic theory is assumed, as in classical statistical mechanics.

It is easier to say what statistical causation isn't, than to say what it is. Causation is not mere correlation. The falling barometer does not cause the rain. They are effects of a common cause. There is a standard way to test for such correlations; they disappear when you control for the common cause. If the pre-existing meteorological conditions are fixed and we manipulate the barometer, we no longer find the state of the barometer to be statistically relevant to rain.

The most popular approach to statistical causation has been to try to turn this methodology into a definition. Statistical causation is positive statistical relevance which does not disappear when we control for all relevant pre-existing conditions.

There are various levels on which one could try to make this idea precise. One could assume a probability measure representing the *physical probability* or *chance* resulting from fixing all the background factors relevant to the target effect excepting the putative cause. Or one could work on the level of the *degrees of belief* resulting from the raw data, which reflect averaging over various possible constellations of background factors.

I. J. Good (1961–62) takes the first approach. He assumes the relevant physical probability measure, reflecting both the pre-existing circumstances and the true laws of nature. In terms of this measure

Brian Skyrms and William L. Harper (eds.), Causation, Chance, and Credence, Vol. I, ix–xii.

qualitative probabilistic causation is just positive statistical relevance. Good offers more – a quantitative explication of the tendency of C to cause E. This is a chance version of sufficient condition, the weight of evidence against C if E does not happen: log [CHANCE (not-E|not-C)/CHANCE (not-E|C)]. It is, of course, also possible to give the comparable chance analogue of a necessary condition version of causation within this framework, i.e. the weight of evidence for E if C does happen.

Reichenbach, Granger, Suppes, Salmon and others take more or less the second approach. They start with probabilities derived from the raw data, and partition to eliminate spurious causation. There are two main strategies. One is to identify an "adequate" partition which covers all the relevant background factors, and to use probability conditional on this partition as the probability where statistical relevance implies probabilistic causation. This strategy can be thought of as an extension of Good's approach, where Good's "physical probabilities" are construed as probabilities conditional on the partition. Specifying the appropriate partition may depend on the principle of temporal priority of causes (Granger), on the concept of state in the relevant physical theory, and on pragmatic factors relating to the application of the resulting concept of statistical causation.

An alternative approach to specifying the appropriate partition is to quantify over partitions. Reichenbach (1956) says that C is causally relevant to E if in the gross statistics, $\Pr(E|C) > \Pr(E)$ and there is no set of events earlier or simultaneous with C such that conditional on these events E and C are probabilistically independent. Suppes (1970) makes a related proposal where the second clause is that there is no partition of events earlier that C such that conditionally on each element of the partition, C and E are probabilistically independent. These proposals cannot be quite correct as they stand. Averaging can create spurious *independence* as well as spurious dependence in the gross statistics (see Simpson 1951). Then we can have a genuine case of statistical causation which these definitions miss because there is no dependence in the gross statistics.

One might try a more elaborate version of the quantifying approach: C is a probabilistic cause of E if there is a P-partition such that for every P-partition which refines it the probability of E on C conditional on the partition remains the same and $\Pr(E|C) > \Pr(E)$ conditional on each member of the partition. Here P-partitions are ones generated by events prior to C, but a more general account which makes contact with ergodic theory is possible (by generalizing P-partitions to σ-algebras of invariant sets). This version of the quantifying approach, however, amounts to a way of identifying an adequate partition and proceeding as before.

Indeed, although philosophers often write as if the appropriate partitions are finite or countable, in many contexts they are not and heavier mathematical machinery is required. Probability conditional on a partition gives way to probability conditional on a σ-algebra. Generalizations of de Finetti's theorem and ergodic theory become relevant tools for identifying the appropriate physical probabilities. Thus the last group of papers in the volume is relevant to the mathematical bridge between two approaches to probabilistic causation discussed in the first group of papers: that which takes physical probability as given and that which starts with degrees of belief.

Once a general notion of statistical causation between event types – or more generally random variables – is in hand, questions arise as to its correct application. Any theory of rational decision which evaluates actions in terms of their consequences must come to terms with statistical causation. Spurious correlation can cause problems in Savage's statistical decision theory if the states are not chosen with sufficient care (Blyth 1972, and the following discussion especially Pratt's remarks on independence). And related difficulties arise in Jeffrey's decision theory, where the remedy of careful choice of states is not available. (Gibbard and Harper 1980, Cartwright 1979)

Causal judgements are important for prediction and for belief revision. And at least one important form of scientific explanation explains events in terms of their causal antecedents. Proper control for spurious correlation is also a concern for these areas of application (e.g. Salmon 1980). In addition, it is sometimes important to consider causal chains of actual events. In classical contexts it is often natural to use principles of locality of causation, identification of causal chains with potential signals, and requirement of energy-momentum transfer for causation as aids in constructing the appropriate chains. The Einstein–Podolsky–Rosen gedankenexperiment in quantum theory derives much of its air of paradox from the fact that it is a case where these principles no longer cluster together with statistical causation (Bell 1964). The papers on the application of statistical causation address a tangle of metaphysical and epistemological questions which is central to all of science.

BRIAN SKYRMS *University of California at Irvine*

REFERENCES

Bell, J. S. (1964) 'On the Einstein–Podolsky–Rosen paradox', *Physics* **1**, 195–200.

Blyth, C. R. (1972) 'On Simpson's paradox and the sure-thing principle', *Journal of the American Statistical Association* **67**, 364–366. (See also commentary by Good, Lindley, and Pratt.)

Cartwright, N. (1979) 'Causal laws and effective strategies', *Noûs* **13**, 419–37.

Gibbard, A. and Harper, W. (1980) 'Counterfactuals and two kinds of expected utility', in *Ifs*, Harper *et al.* (eds.) (Reidel: Dordrecht).

Good, I. J. (1961–62) 'A causal calculus', *British Journal for Philosophy of Science* **11**, 305–328; **12**, 43–51; 13, 88. Reprinted as Chapter 21 in *Good Thinking* (University of Minnesota Press: Minneapolis, 1983).

Granger, C. W. J. (1969) 'Investigating causal relations by econometric models and cross spectral methods', *Econometrica* **37**, 424–438.

Jeffrey, R. (1965) *The Logic of Decision* (McGraw Hill: New York); 2nd rev. ed. (University of Chicago Press: Chicago, 1983).

Reichenbach, H. (1956) *The Direction of Time* (University of California Press: Berkeley and Los Angeles).

Salmon, W. C. (1984) *Scientific Explanation and the Causal Structure of the World* (Princeton University Press: Princeton, N.J.).

Simpson, E. H. (1951) 'The interpretation of interaction in contingency tables' *Journal of the Royal Statistical Society Ser. B* **13**, 238–241.

Sims, C. A. (1972) 'Money, income and causality' *American Economic Review* **62**, 540–562.

Suppes, P. (1970) *A Probabilistic Theory of Causality* (*Acta Philosophica Fennica*, Vol. 24) (North Holland: Amsterdam).

PART A

ON THE NATURE OF PROBABILISTIC CAUSATION

C. W. J. GRANGER

CAUSALITY TESTING IN A DECISION SCIENCE

1. DECISIONS

1.1. Economists and workers in other decision sciences frequently use words such as "causal" or "cause" in their writings, as a glance at the listing of titles in the Social Science Citation Index will easily confirm. A number of papers have been written specifically on these concepts although few formal definitions have been attempted.[1] In Economics, we observe values for our major variables each month, or some similar regular time-period, and because of this it can be claimed that there is some generating mechanism which produces these values. One of the main tasks of the economist is to try to untangle and understand this mechanism and in the search for the fundamentals of this process, the concepts of "law" and "causation" arise very naturally. However, the search for causality may be motivated by more than just academic interest, as will be discussed below.

1.2. Observed economic activity is almost invariably the result of decisions made by economic agents, such as individual consumers, households, firms, institutions, and local governments. The structure and properties of these decisions plus their consequences forms the subject of microeconomics. Aggregation of the outcomes of the decisions give rise to the macroeconomic variables, such as gross production, imports, employment, and price indices, although some macro-variables can be thought of as outcomes of decisions by the central government, such as tax rates, government expenditures, deficits and parts of money supply.

The fact that decisions are at the core of the economy enhances the importance of causal ideas, suggests sources for causal relationships and places limitations on the form of any causation. It also allows a particular interpretation of the outcome of any causal analysis, such as a statistical test, as will be seen.

1.3. In its simplest characterization, a decision can be taken to consist of three components, (a) a decision rule, (b) a set of current or

[1] Additional comments and references can be found in the notes at the end of the paper.

Brian Skyrms and William L. Harper (eds.), Causation, Chance, and Credence, Vol. I, 3—21.

recent values of relevant variables to be inserted into the rule, and (c) a decision outcome.

As a simple example, if one is interested in buying some grapes in a store, the decision rule (based on a maximization of expected utility) might be to buy a pound of grapes if the price is less than $1 a pound. The input to the decision will be price (potentially observable) plus the appearance or apparent quality of the grapes in the store (an unobserved cause). The observed outcome is the amount purchased, either zero or approximately a pound if the price is correct. The actual amount purchased may not be exactly one pound, despite the decision rule, because of practical considerations.

(a) The decision rule relates the set of present circumstances, or information, and converts them into a proposed decision. It is essentially a mapping from the inputs to the outcome. It will be based on beliefs about the benefits arising from the decision and thus about the form of the generating process for that section of the economy relevant to the decision. It will also use some criterion that is being applied to choose between alternative possible rules. Whether or not the rule leads to "optimal" or even "sensible" decisions is outside the scope of this discussion, and is not relevant for causality, what matters is that some rule is applied.

The rule that is used will be based on accumulated experience about the economy or industry or region plus observations of available data, possibly over a long period and various expectations about the relevant sector of the economy. These expectations will partly depend on beliefs about the generating mechanism. For convenience, it will be assumed that the degree to which a decision-maker believes some proposition, such as the importance of some variable in the mechanism, can be expressed in terms of a single probability. This will be a subjective probability based both on direct evidence, if available, experience and intuition. The rule will thus incorporate both the believed economic relationships and the degree to which this belief is held, which will partially measure uncertainty or risk in the decision process. In the example just given, the uncertainty could be about the health benefits from eating grapes.

(b) The inputs to the rule will be values, or estimates, of variables required to implement the rule. It is important to note that these values can only consist of those available at the time the decision is made, although they can include expectations of future values.

(c) The outcome of the decision will usually have an impact on some real part of the economy, such as production, a change in employment, an investment or a purchase. However, there can be a difference between the decided outcome and the actual outcome, due to the presence of uncertainty in the outcome implementation process.

1.4. With a fixed decision rule the decision process can be given a simple causal interpretation, the inputs to the rule can be thought of as causes and the outputs as effects or caused variables. Part of the inputs may be observable to an outsider and part of the outcome may be observable. There is thus potentially an analysis, such as a statistical test, that can be applied to the observable parts of the inputs and the outputs. Because there will be unobservable parts, any "causality" so detected will be probabilistic in nature, so that rarely will an observed input inevitably lead to a particular value of the output.

1.5. The decision sequence, as characterized here, takes the form (i) collection of the values of the observed variables, (ii) these values are placed into the decision rule, (iii) the decision outcome occurs, (iv) implementation of the decision outcome, (v) observation of part of this implementation. There is almost inevitably a time delay between the observed input (the cause) and the observed output (the effect). This delay may be small in time compared to the interval over which data is collected for the economy. The possibility of "instantaneous" causality will be discussed in Section 3 of this paper.

1.6. With this formulation, the decision rule is the "causal mechanism." However, as this rule is itself based on beliefs it can change through time as beliefs change, unlike a physical law. In particular, if the degree to which a belief is held changes, so the decision rule may change. Thus, the outcome of a causal analysis may change the degree of belief and thus the "behavior" (that is the decision rule plus outcome) of the economic agent. (C.f. Figure 1.)

In the grape purchase example, a new study may indicate that eating grapes reduces cholesterol, thus changing the belief about the health benefits of the purchase and an increase in expected utility. The new rule might become – buy grapes if the price is under $1.10 per pound.

1.7. As a further example, consider a small manufacturing company, whose owner decides every Friday evening the production schedule for the coming week. The total quantity to be produced will depend on current inventory levels of the product and of raw materials, plus expectations of immediate demand for the product. These

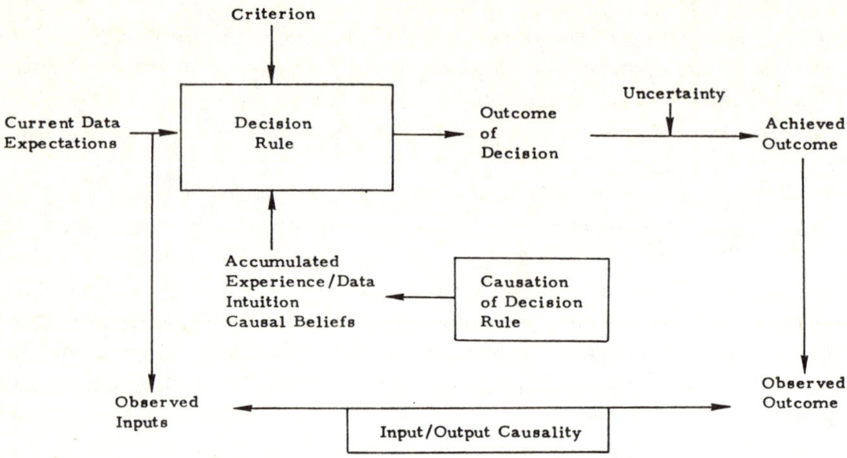

Fig. 1.

expectations can be functions of beliefs about the present and future state of the local economy, likely pricing behavior of competitors and a belief by the owner that when the stock market index rises 10% then a few days later demand for his product will increase by between 3% and 6%. However, he admits that he is not totally convinced in his belief of the influence of the stock market on his product, and holds a degree of belief of 0.7. Part of the accumulated experience will relate previous expectations of demand with actual demand, so learning will have occurred. From these beliefs, and others, a decision rule can be constructed which will give the production level that will maximize the certainty equivalent profit for the week. The inputs will include his prices, assumed to be predetermined, number of production workers, inventory level, stock market price index and possibly a group of purchased forecasts of the economy. The output will be the production schedule for the coming week. The observed outcome will be the actual production achieved during the week, which may differ from the planned output because of unexpected power outages, worker illness, machine breakdowns, and so forth. In this example there will be a delay of at least a week between the observed inputs, such as the stock market prices, and the actual production.

Because of the various unobserved components, there will be, at best, a probabilistic relationship between the inputs and the outputs, which have been equated with causes and effects. This might be thought of as either an inherent stochasticity in the situation or the impact of unobserved causes. From the data available to an outsider, it will usually not be possible to determine the actual decision rule being used. It may also be argued that there is a continuum of causes through time, from the input variables to the observed output, most of which are unobserved. An example would be the electrical impulses within the owner's brain which enable him to apply the rule. The existence or not of unobserved variables is of little or no importance in a statistical analysis if one knows *nothing* about the properties of unobservables and so discussion of such questions is outside the province of the statistician, although they can be of considerable importance to a philosopher.

1.8. The analysis so far suggests causal implications for two different parts of the decision mechanism. For a fixed decision rule, the (observed) inputs may be thought of as causes for the observed outputs. In this case, inventory level will be a cause of observed production.

However, when forming the decision rule, various beliefs about causes may also be utilized, such as the causation from stock market prices to customer demand in the example. It is essential that these two causal pairs are kept distinct in the analysis. To identify these two causations, the first will be called "input/output causation," the second "causation of the decision rule." It should be noted that causation of the decision rule will typically be the result of aggregation of input/output causes by other decision rules and will have an associated value for the degree of belief about this cause.

1.9. Suppose now that an analysis is conducted of the causal relationship between stock market prices and the production by the industry to which this company belongs. If the owner believes that a reasonable definition of causality has been used, sensible data were utilized and the statistical analysis was sound, then the results of this analysis may alter the degree of his belief about the relationship between the stock market and demand for his own product. The study need not be exactly on his own data or be perfect in every way for it to change his degree of belief from 0.7 to 0.8, say. This change may alter his behavior, that is it will change his decision rule.

It can be argued that a great deal of intellectual, experimental and statistical analysis is undertaken in order to try to change the degree of belief of one's colleagues or the public at large. In many fields, one would not expect to completely convince the profession that one is correct, but one can hope to influence them towards believing in the correctness of one's position.

The studies of causation may thus influence the causation of the decision rule and a change in the rule may then influence the input/output causation. For example, if it was widely held that smoking cigarettes is not harmful to one's health, then an increase in advertising expenditure or a decrease in price will alter cigarette sales. Advertising will be an input/output cause of sales. A change in the fundamental belief, so that it becomes generally accepted that smoking is harmful, will change the decision rule and thus the effects of advertising and prices on sales. The change in belief can occur from evidence that is suggestive and somewhat convincing but is not completely convincing. In particular, it should be noted that the evidence will be probabilistic in nature, as not every smoker is necessarily harmed.

1.10. Two other examples will be considered briefly. In the first, a consumer considers buying a new automobile. The inputs will be observables such as prices, income, interest rates, plus unobservables such as the expected utility to be derived from buying a particular car and the utility lost from spending the needed amount of cash, the opportunity cost. The first of these utility expectations may include beliefs about the causal relationship between the size of a car and the safety of its passengers. The decision rule might be to choose the car with the largest net expected utility (or perhaps certainty equivalent utility to allow for risks) provided the cost is less than some predetermined amount. No individual will make enough of these decisions to generate enough data to perform a test on input and output variables, but aggregating information from many individuals might find a causal relationship between interest rates and the proportion of large to small cars purchased. A study of the relative safety records of large and small cars could clearly effect the decision rule.

A further example might involve a state government which is considering the introduction of legislation making the wearing of seat belts compulsory for automobile drivers and passengers. The decision of whether or not to introduce such a law will be based on the expected costs in terms of lost freedoms by citizens, and benefits, in terms of lives saved and injuries reduced. The expected benefits will be partially based on the degree of belief about the effect of seat belt wearing on injuries and lives saved. These beliefs will be partly based on the experiences of other states and nations, regional survey data and, possibly, experiments, all of which may be considered as informal causal tests.

1.11. The demand for a better understanding of causal relationships arises from the need to make good decisions and this will depend on a sound understanding of the economy, or those parts of interest to the decision maker. In the first example, the firm's owner will need to understand the determinants of the demand for his product. In a stable economy, he will only need a good forecasting procedure, using some leading indicators, for example. However, if the economy experiences a structural change or enters a region in which there is little past experience, such as high inflation and low nominal interest rates, then a more fundamental belief or understanding about the determination of demand is required. If it is believed that one has identified a true cause, then this relationship should be helpful in the decision process during all periods. That is, a true cause indicates a deeper relationship than a mere leading indicator.

If decision rules are based on causal beliefs, the obvious question arises of how these beliefs can be checked. If causation is not clear from intellectual analysis, the question becomes an empirical one and thus has to be tackled by a statistical analysis. It is also clear that a statistical analysis can only be suggestive rather than provide a complete proof of causation, as it will be based on an incomplete, imperfectly measured, finite set of data. At best, such an analysis can only influence the degree of belief about a causation. It may be more influential if it is also associated with a persuasive theoretical reason for causation but not everyone will require the existence of such a theory. The link between smoking and cancer is an example where degrees of beliefs have been changed solely on statistical evidence plus a widely-held agreement that such a causal link is sensible.

2. AN OPERATIONAL DEFINITION

2.1. To perform a statistical analysis definition of causation is required that has operational implications. A definition that has been widely used in econometrics, although is by no means fully accepted, can be constructed from two necessary conditions for causation:

Basic Axiom 1 (Temporal Priority)
A causal event will occur before the event caused (effect).

Basic Axiom 2 (Special Relationship)
A causal event will contain information about the effect event that is not contained in any other event occurring no later than the cause.

Suppose that x_t, y_t are a pair of (real) random variable sequences observed at discrete time intervals $j = 0, 1, 2, \ldots, t$ and that an event is equated with x_t or y_t taking a value in some specified region of the real line. The axioms can be formally stated as

Causality Definition

y_t causes x_{t+1} if

$$\text{Prob}(x_{t+1} \in A \mid \Omega_t) \neq \text{Prob}(x_{t+1} \in A \mid \Omega_t - y_t) \quad \text{for some set } A \tag{1}$$

where Ω_t is the set of all observable events available up to time t and $\Omega_t - y_t$ is this set minus information exclusively in the sequence y_{t-j}, $j \geq 0$.

A consequence of this definition is:

Forecastability

If any function $h(x_{t+1})$ of x_{t+1} is to be forecast using any cost function (to use as a criterion to choose between competing forecasts) then one is never worse off (in terms of expected cost) using y_{t-j}, $j \leq 0$ and sometimes one is better off. The two basic axioms ensure that causality has a forecasting implication but, of course, causality is not equated merely to forecasting ability, at the level of generality so far considered.

The first step in making this definition operational is to replace Ω_t with a smaller, more practical information set I_t, which is assumed to contain *at least* x_{t-j}, y_{t-j}, $j \geq 0$, but may also contain z_{t-j}, $j \geq 0$, where z_t is some vector of measurable variables not including (or perfectly related to) y_{t-j}. It will naturally contain the deterministic variable t and hence any function of t, including a constant. If now

$$\text{Prob}(x_{t+1} \in A \mid I_t) \neq \text{Prob}(x_{t+1} \in A \mid I_t - y_t) \tag{2}$$

then y_t is said to be a *prima facie* cause of x_{t+1} with respect to the information set I_t. As the question is being asked – does y_t cause x_{t+1} – it must be thought of as a possible cause. Using (2) suggests that it remain a possible, or *prima facie*, cause after passing a test based on conditional probabilities. However, the *prima facie* phrase, which was introduced by Suppes (1970), indicates that if we had used a different, presumably larger, information set y_t might not have passed his more stringent test and so would then have been realized not to be a cause. In almost any statistical analysis the interpretation of results can be affected by missing variables and so this problem is by no means new or unique to causality questions. Using the limited but realistic I_t rather than the unachievable, and possibly even undefinable Ω_t, means that the possibility of an untrue "cause" being found is always present. However, if tests are based on increasingly larger information sets containing all sensible other explanatory variables, then if y_t passes the test based on (2) then one may have increasing confidence, or degree of belief, that y_t does cause x_{t+1}.

2.2. There are, in fact, very few powerful statistical procedures based on whole distributions, as required by (2). There is therefore some point in considering yet more limited, but more useable, definitions. Rather than consider the whole conditional distribution, one can concentrate on the conditional mean. The definition then becomes:

$$E[x_{t+1}|I_t] \neq E[x_{t+1}|I_t - y_t] \tag{3}$$

then y_t is a "*prima facie* cause of x_{t+1} in mean with respect to the information set I_t." It follows immediately that the mean squared error of forecast residuals is less if y_{t-j}, $j \geq 0$, is used to help forecast x_{t+1} than if the y's are not used. Most of the actual statistical tests of "causation" in the economics literature are based on this definition, or equivalent statements, when it is called tests for Granger or Weiner–Granger causation to distinguish it from other possible definitions. Clearly, as the definitions become more operational they become less general and so the possibilities for incorrect interpretation increases.

2.3. There are a number of immediate consequences of the various definitions, including:

(a) If y_t causes x_{t+1} then x_t may but need not cause y_{t+1}. The basic axiom 1 is seen to be essential for this non-necessary symmetry.

(b) If x_{t+1} is self-deterministic, so that

$$\text{Prob}(x_{t+1} \in A | x_{t-j}, t) = 0 \text{ or } 1 \text{ for every } A$$

then y_t cannot cause x_{t+1} because it cannot add any useful information. In particular, if the forecast of x_{t+1} based on x_{t-j}, $j \geq 0$ and t gives forecast residuals with zero variance then clearly adding es cannot reduce this variance. It follows that a linear :annot be tested to cause another linear trend using these ..ons.

(c) The definitions cannot be used to test causation of a unique event, as no statistical test can be devised for use with a sample sized one.
(d) If y_t causes x_{t+1} and x_{t+1} causes z_{t+2} then y_t may, but need not, cause z_{t+2}, as some simple examples show. The non-sequential causation can occur because y can cause part of x but the other part of x causes z.
(e) If variables x_t, y_t are replaced by any backward (linear) filters applied to the series individually, then the causation conclusions remain unaltered. So, if y_t causes x_{t+1}, then $y_t - ay_{t-1}$ will cause $x_{t+1} - bx_t$, for example. However, if two-sided filters are applied, involving both past and future values, then causation conclusions can be altered, because Axiom 1 is no longer applicable. This problem can occur with some seasonally adjusted data in economics.
(f) Using (1) or (2) and Bayes theorem gives

$$\text{Prob}(y_t \in C | x_{t+1} \in A, I_t') \neq \text{Prob}(y_t \in C | I_t') \tag{4}$$

where $I_t' = I_t - y_t$. Thus, if (2) gives a probabilistic form of "necessity," in that if $y_t \in C$ is observed this effects the probability that $x_{t+1} \in A$, then (4) gives a form of "sufficiency," in that if we observe that $x_{t+1} \in A$ it suggests that in the earlier period the problem of $y_t \in C$ is changed. It is seen that these definitions provide simple generalization of the necessity and sufficiency concepts that have been connected with causality by philosophers for a couple of thousand years.

2.4. There are of course a number of practical difficulties in applying the definitions. The most important, the possibility of missing variables, has already been mentioned. A particular aspect of this problem is if the variables are measured with errors having temporal structure, which can lead to misleading result from the test. If a trio of variables x, y, z have causal relationships, then in some instances causality from x to y, say, can be disrupted if z is controlled. A further problem arises if the true causal lag is, say δ time intervals, which is small compared to the actual time interval for which economic data is recorded. The effect of this kind of temporal aggregation is to lose some causal relations, and confuse some two-way causal relationships. This question is considered further in Section 3 below.

2.5. A special problem concerns the time at which a variable occurs compared to the time at which it is observed. It typically takes some considerable effort to gather together the information required to estimate a macroeconomic variable, such as gross national product, total unemployment or the level of imports. Thus, the announced figure for the rate of unemployment for November may not be available until January. In this case, the figure presented in January is known to relate to November and is thus designated accordingly. On some occasions, time of observation is not the time of the actual occurrence but the actual time at which the event occurred is not known and so a proper temporal designation cannot be made and this can lead to incorrect interpretation of tests.

The obvious example concerns lightning and thunder. Lightning is always observed before thunder and so if these events are placed in time when observed then the definitions will suggest that lightning causes thunder, but if both events are placed at the time they first occur, they will be found to be jointly caused by an unobserved event, being an electric arc jumping through space. Once this possibility is realized it becomes somewhat less of a problem.

Of course, the extreme example of not knowing when something occurs is with an unobserved variable. In particular, in an equation such as

$$x_t = by_{t-1} + e_t$$

the residual e_t is typically ot directly observed and it is given the temporal designation t from convention and because that is the first time it can be derived from the equation. Of course the unobserved input which is being proxied by e_t could have actually occurred earlier.

3. SIMULTANEITY AND INSTANTANEOUS CAUSALITY

3.1. If I_t is an information set consisting of x_{t-j}, y_{t-j}, $j \geq 0$ plus possibly other variables, it is frequently found that the conditional distribution of the vector (x_{t+1}, y_{t+1}) given I_t has components that are not independent. In particular, if a (linear) model is constructed relating each of x_{t+1} and y_{t+1} to the components of I_t, the two residuals are often found to be correlated. The system may be said to display simultaneity. The immediate question is – why do we observe simultaneity? This might be phrased as – what is the cause of the simultaneity, if it is believed that every observed phenomenon must have a cause?

There are three obvious answers to the question:

(a) there is a missing common cause,
(b) one residual causes the other but this is hidden by temporal aggregation, or
(c) there is true instantaneous causality.

In (a), (b) "causality" is defined as (1) or (2), whereas (c) concerns the possible causation of x_{t+1} by y_{t+1} or *vice versa*, however that might be defined.

3.2. To illustrate these three cases, consider a pair of series x_t, y_t observed at equal intervals, $t = 0, 1, 2, \ldots$ which are completely temporally uncorrelated, so that

$$\operatorname{corr}(x_t, x_{t-s}) = 0 \qquad s \neq 0$$
$$\operatorname{corr}(y_t, y_{t-s}) = 0 \qquad s \neq 0$$
$$\operatorname{corr}(x_t, y_{t-s}) = 0 \qquad s \neq 0.$$

However, simultaneity may be present, so that $\operatorname{corr}(x_t, y_t) = \rho \neq 0$. Thus, with definition (3), concerning causality in mean, neither series causes the other. If there exists a third series w_t so that

$$x_{t+1} = \beta_x w_t + \varepsilon_{xt}$$

and

$$y_{t+1} = \beta_y w_t + \varepsilon_{yt}$$

with ε_{xt}, ε_{yt} uncorrelated with each other and with w_t, then obviously

$$\rho = \frac{\beta_x \beta_y \cdot \operatorname{var} w}{\sqrt{\operatorname{var} x \cdot \operatorname{var} y}}. \tag{5}$$

Thus, if w_t exists as a common cause but is unobserved, it can explain the simultaneity. As explained earlier in 2.5, it is unclear when an unobserved variable occurs or even exactly what it is. One possibility is that $w_t = x_{t-\delta}$, where $0 < \delta < 1$ so that there is causation of y_t by $x_{t-\delta}$, which is unobserved and also the x_t series would display autocorrelation if it were observed more frequently. There is little reason to believe that the data in economics is gathered, or observed, with the optimal sampling interval. Thus, cases (a) and (b) can be explained by missing causal variables using causal as defined in 2.1. It should be noted in (5) that both β_x and β_y need to be non-zero for simultaneity to occur.

In case (c) one can have $\rho \neq 0$, that is simultaneity, even if no unobserved w_t exists, so that $\text{corr}(y_t, x_{t-\delta}) = 0$ all $\delta \neq 0$ and there is no other w_t with $\beta_x\beta_y \neq 0$.

3.3. If it is agreed that simultaneity occurs for some reason, which is another way of saying that it has a causal explanation, and if it is agreed that either (a) or (b) of (3.1) are the only possible explanations, then there is the important conclusion that no new definition of causation is needed to investigate simultaneity. However, if (c) is possibility, the analysis becomes a great deal more complicated. I believe it can be argued that it is extremely unlikely that instantaneous causality can occur in a decision science. Arguments in favor of this proposition are presented in Section 3.4 and suggestions for analysis of cases (a), (b) are then discussed in 3.5.

3.4. It is convenient to start with a new basic axiom:

> *Axiom 3:* A cause cannot occur after the effect that it is causing.

This axiom is generally accepted, although of course not universally, and will be assumed to be true in what follows. Temporal ordering of events will be taken to be as perceived by an individual and it will be assumed that all individuals have the same perception of this ordering for any events under discussion. The Basic Axiom 1, in 2.1, assumes cause occurs before effect, giving a temporal ordering basis for causality. Instantaneous causality can exist only on the knife-edge in time between causality as in Axiom 1 and the impossibility of the temporal reverse ordering excluded by Axiom 3. A change in the cause variable will immediately and with no time delay of any size (or at least detectable by an available measurement instrument) change the probability of the caused variable occurring in some region.

Because of the way certain types of variables are generated, the strict application of Axiom 3 excludes instantaneous causation in some cases. A variable may be called "instantaneously recorded" or IR if it can be recorded at every instant of time, at least conceptually. Examples are prices, employment, level of wealth and temperature. A variable that is the accumulation of a quantity over a discrete interval of time may be called an "accumulated" or A variable. Examples are production, exports and rainfall. From Axiom 3 it is clear that an accumulated variable cannot instantaneously cause another accumulated variable as part of the first variable will inevitably have occurred before part of the second. Similarly, an IR variable cannot instantaneously cause an A variable because it will have occurred after most of the A variable has physically occurred (as opposed to when it is observed). Thus, instantaneous causality can only occur from IR to IR or from A to IR variables. With a causality definition based on temporal ordering, such restrictions do not arise.

The central importance of decisions in economics also raises doubts about the practicality of instantaneous causality. Most decisions are made within the brain and most brains are probably single-channeled, so that several decisions cannot be made simultaneously. Thus, if a single decision maker is concerned with several decision situations, a change in a set of input variables cannot simultaneously lead to changes in the probabilities of all the output variables. Further, the process of collecting input data and applying a rule will usually result in a noticeable delay between the observable input and the observable output, as noted in Section 1.5. To react always very quickly to new information may lead to unnecessarily high costs for the decision maker. Consider the example of 1.1. involving a prodution scheduling decision. The example has the owner making the decision weekly. If he reacts to every piece of information relating to demand for his product, the schedule could be changed frequently, with probable extra costs in worker compensation, raw material deliveries and because of changes in machine maintenance schedules. Similarly, a consumer is not necessarily better off changing his purchasing plans every time any price is observed to change, as he will end up doing nothing but make plans and will never actually buy, or consume, anything. Thus, in most cases, delays in decision making correspond to good economic practice. An exception might be the speed of reaction to an enormously important piece of news, such as a competing firm going bankrupt, a president being shot or war being declared. As these are necessarily very unusual events, or outliers to a statistician, they will not be easily incorporated into a statistical test and so do not fall within the framework being discussed here.

These arguments do not completely exclude (effectively) instantaneous causality in economics but do suggest that it is unlikely.

3.5. In some parts of economic literature, the phrase "instantaneous causality" has been associated with the extra "forecastability" of x_{t+1} if y_{t+1} is added in the information set I_t. If there is simultaneity present, so that $\rho \neq 0$ in the example in 3.2, then one will always find instantaneous causation of x_{t+1} by y_{t+1} and also of y_{t+1} by x_{t+1}, so this is seen to be not a useful definition. Basman (1965) expands on this theme in a convincing fashion.

3.6. If simultaneity is observed, possibly after conditioning out a group of other random variables observed no later than the pair of variables of interest, this simultaneity deserves an explanation and the possibility of causation from one variable to the other has to be discussed. It is clear that a causal interpretation cannot generally be made solely from the information in the data. As a simple example one can consider a pair of variables x_t, y_t that are jointly white noise, have a Gaussian distribution and are independent of all variables in some vector z_t, which together with x and y makes up the available information set. If x_t, y_t have a non-zero correlation ρ, then any measure of the extent to which x can be explained by y, such as R^2, will be identical to the extent to which y can be explained by x. Such symmetries do not necessarily occur for non-Gaussian variables, although the R^2 measure remains symmetric.

As Axiom 1 is not available some other information, exterior to the data, is required to break the impass. For example:

(a) It may be "known," or at least strongly believed, that x_t is exogenous to the system, so that y cannot cause x and a non-zero ρ value would then be suggestive of causation from x to y. An example is air-temperature outside a house causing changes in the temperature inside the house if an instantaneously reacting thermostat is installed. Another example might be that part of a control variable, such as money supply, that is not explained by the government's reaction function. As it is controlled, this part may be considered exogenous. The source of this knowledge or belief is important, it may be some highly respected scientist or economist or arise from an act of faith, via statements made by a priest, for example. Although some individual may find this knowledge convincing and be prepared to base causal beliefs on it, this belief may well not be transferable to other decision makers.

(b) There may be a convincing theory about why y causes x but no corresponding theory in the other direction. Of course, what theory is considered satisfactory and what is unsatisfactory is subjective and so again beliefs will not necessarily be transferable from one decision maker to another.

(c) It might be considered unlikely that a "small event" can cause a "big event" and so any causation must be in the opposite direction. For example, a relationship may be observed between an index of sunspot activity and an economic variable, after conditioning out other variables. As it is unlikely that the state of the economy affects the sun, the only possible causation is the opposite direction. However, general use of this principle would be unjustified, as some small events can effect large ones, "the straw that breaks the camel's back" being the traditional statement.

(d) A somewhat similar belief is that a slowly changing sequence can affect a rapidly changing one but that the reverse direction is less likely. For example, a business cycle can influence the amplitude of a seasonal component but not *vice versa*. A further example from earlier sections of this paper is that changes in causal beliefs used in the decision rule will induce changes in input-output causation but the reverse will not be detectable.

Doubtless other methods of introducing structure to the simultaneous situation are also possible. A great deal of the economic causal literature considers this situation, such as the Wold–Simon causation discussions (see Blalock 1961) and the related path-analysis due to Sewell Wright. They typically use assumptions such as (a) and (b) but the theory is generalized to finding causal relationships between groups of variables, on the assumption that other causations do not occur. Any tests based on these ideas are thus useful only conditional on the correctness of the non-causality assumptions that are being made. If one believes that the assumption is correct the test may be individually useful. However, it is usually quite impossible to test the assumptions made within the simultaneous framework. It is my strong belief that this creates an important weakness in these tests compared to those based on the temporal ordering Axiom 1. These tests are, of course, trying to solve the rather difficult and particularly intractable problem of how to untangle simultaneous relationships.

3.7. A lot of attention is paid to the question of how to give a causal interpretation to an observed relationship but little thought is given to explaining lack of causation when there is no apparent relationship.

The basic difficulty in this situation can be illustrated by a situation in which a pair of variables X, Y are jointly white noise, have a Gaussian distribution and are independent, so that $\rho = 0$. If a pair of random variables are independent it will probably be generally agreed that there can be no causation between them. If now a third variable, Z, is introduced, it can be true that within the triple X, Y, Z the pair X, Y are no longer independent, in that there can be a non-zero conditional correlation $\mathrm{corr}(X, Y/Z)$ and so it becomes natural to search for a causal explanation. Thus, even independence for a pair of variables can be lost as the information set is expanded. Incidentally, some writers have been critical of the use of the words "cause" or "causal" with respect to statistical testing, but no similar criticism seems to have been applied to the words "independent" or "independence," which are similarly value-loaded.

Department of Economics
University of California, San Diego
La Jolla, CA 92093

NOTES ON SECTIONS OF THE PAPER

1.1 The opinions expressed in this paper are entirely my own. I would not presume to try to speak for all decision scientists, or even all economists. Naturally, causality is a controversial topic in these fields, with no consensus of opinions. The main definition discussed in this paper has been found useful in various fields, such as history (Smith 1984) and political science (Freeman 1983). For some alternative views, see Zellner (1979, who espouses the views of the philosopher Feigl). Hicks (1979) and Simon (1970). The link between the definition I support and some writings by philosophers and others is discussed by Spohn (1984). Because of my background, most of the examples are from economics. An early, influential application was by Sims (1972).

When statisticians consider causality, there are essentially two basic types. In cross-sectional causality, members of a population are examined and we are asked to explain why they lie in different places on the distribution of the variable of interest – such as, why does family A use more electricity than family B. In temporal causation, one asks why the parameters of such a distribution, such as the mean or variance, have changed through time, so one could ask why electricity demand is less this year than last. This paper discusses only temporal causality.

1.2 The basic point in this section was made by Hicks (1979).

1.3 In theory, if a decision rule is time-invariant the identical inputs should lead to the same decision. In practice, this idea is of little relevance as circumstances can never be identical, as the fact that a previous decision was made, and something learned from it, means that things have changed.

[1.9] One clear difference between the two types of causality is in the lags involved. For most decisions, the delay between observed input and observed output may be small, but the lag between the appearance of a causal analysis and changes in a decision rule may be years or decades.
[2.1] The two basic axioms are proposed with the hope that they are generally acceptable as necessary conditions for causation. Of course, they will not be universally accepted. If one has problems with the direction of time, for example, then the first axiom will be unacceptable. Whether these axioms give sufficient conditions for causation will be a personal decision.
[2.2] The definition introduced here is that proposed by Wiener (1956) and expanded in Granger (1969). Papers discussing the actual statistical tests based on this definition include Pierce and Haugh (1977), Geweke (1984) and Nelson and Schwert (1982).
[2.3] An expanded version of this section can be found in Granger (1980). Concerning unique events, an example is the question "what caused the fall of Rome," assuming we agree it did fall. The fact that some other event occurred earlier, such as the growth of Christianity in the city makes this a possible cause but a statistical analysis would be quite inappropriate. If one could observe a number of such "unique events," such as the start of a population problem on Pacific Islands and if this start was always (or usually) preceded by the introduction of Christianity on the islands, a purely statistical case would start to build. However, the case would be much strengthened if all other possible causes were first conditioned out of the analysis.
[3.1] Physicists also struggle with problems of instantaneous causality, see d'Espagnat (1981, 1984).

REFERENCES

Basman, R. L. (1965) 'A note on the statistical testability of "explicit causal chains" against the class of "interdependent" models', *J. American Statistical Association* **80**, 1080–1095.

d'Espagnat, B. (1981) 'The concepts of influences and of attributes as seen in connection with Bell's theorem', *Foundations of Physics* **1**, 205–234.

d'Espagnat, B. (1984) 'Nonseparability and the tentative descriptions of reality', *Physics Reports* **110**, 201–264.

Freeman, J. R. (1983) 'Granger causality and the time series analysis of political relationships', *Am. J. of Political Science* **27**, 327–358.

Geweke, S. (1984) 'Inference and causality in economic time series models', Chapter 19 of *Handbook of Econometrics*, Volume II, edited by Z. Grilliches and M. D. Intriligator (North Holland: Amsterdam).

Granger, C. W. J. (1969) 'Investigating causal relationships by econometric models and cross-spectral methods', *Econometrica* **37**, 424–438.

Granger, C. W. J. (1980) 'Testing for causality: A personal viewpoint', *J. of Economic Dynamics and Control* **2**, 329–352.

Hicks, Sir John (1979) *Causality in Economics* (Basic Books: New York).

Nelson, C. R. and G. W. Schwert (1982) 'Tests for predictive relationships between time series variables: A Monte Carlo investigation', *J. American Statistical Assoc.* **77**, 11–18.

Pierce, D. A. and L. D. Haugh (1977) 'Causality in temporal systems, characterizations and survey', *J. of Econometrics* **5**, 265–293.

Simon, J. (1970) 'The concept of causality in economics', *Kyklos* **23**, 226–254.

Sims, C. A. (1972) 'Money, income and causality', *American Economic Review* **62**, 540–552.

Smith, P. H. (1984) 'Time as a historical construct', *Historical Methods* **17**, 183–191.

Spohn, W. (1984) 'Probabilistic causality: From Hume via Suppes to Granger', pp. 64–87 in *Causalita e modelli probabilistici*, M. C. Galavotti and G. Gambetta (eds.) (CLUEB Editrice: Bologna).

Suppes, P. (1970) *A Probabilistic Theory of Causality* (North Holland: Amsterdam).

Wiener, N. (1956) 'The theory of prediction', *Modern Mathematics for Engineers*, series 1, E. F. Beckenbeck (ed.), chapter 8.

Zellner, A. (1979) 'Causality and econometrics', *Carnegie-Rochester Conference on Public Policy*, vol. 10, pp. 4–56, K. Brunner and A. Meltzer (eds.).

I. J. GOOD

CAUSAL TENDENCY: A REVIEW[1]

1. INTRODUCTION

We can judge of the perfection to which a science has come by the facility, more or less great, with which it may be approached by calculation.

Quetelet (1828), as quoted by Landau & Lazarsfeld in the *International Encyclopedia of Statistics*, p. 828.

Suppose that some event F occurs and later an event E either occurs or does not.[2] I am going to talk about *the extent to which F tends to cause E*, or *the contribution from F to the tendency for E to occur*, or *the tendency of F to cause E*, denoted by $Q(E:F)$. For example, F might be the event that the captain of a firing squad shouts "Fire" and E might be that the victim's life is Ended (F for Fire, E for Ended). The negation of E is denoted by $\bar{E}$, which in this example means that the victim does not drop dead. The question arises whether $Q(E:F)$ can be sensibly defined in terms of physical probability = material probability.[3] I believe there is a good definition which, to state it roughly, is the *weight of evidence against F if E does not occur* (Good 1961, 1980e). If the victim does not drop dead it is strong evidence that the captain did not shout "Fire". In other words the captain's order to fire had a great deal to do with the potential death of the victim.

My plan is to explain this definition, sharpen it, defend it, and apply it. The approach will be basically an example of *constructive quantitative applicable philosophy*. Or it could be called semi-quantitative rather than quantitative because of the difficulty of measuring probabilities with precision. The desideratum–explicatum method will be used, but with proofs omitted because this paper is largely a review. The desiderata are less convincing than those that have sometimes been used for subjective or logical probability,[4] amount of information,[5] weight of evidence,[6] or explicativity,[7] but they are the best and simplest that I found to lead to a sensible explicatum in a year of evenings and weekend work in 1959. Because the desiderata are not totally convincing in themselves it is especially worthwhile to examine

Brian Skyrms and William L. Harper (eds.), Causation, Chance, and Credence, Vol. I, 23–50.

applications of the explicatum of causal propensity to see if they seem to make sense and this is one reason why I shall give several examples.

At first it seems that probabilistic causality can exist only in an indeterministic world (Einstein's name for the universe), but one can wriggle out of this conclusion in the following manner. Suppose that the world is deterministic but that we are considering some phenomenon, such as boiling a kettle of water, that is impracticable or impossible to describe in complete detail. Then it is *as if* we were dealing with an indeterministic system. This observation applies also to criminal trials. In a deterministic world, "in principle" *tout comprendre c'est tout pardonner* to quote Madame de Staël (see Bartlett 1980, p. 417). Even if the real world is deterministic, we need to assign credit and blame, and perhaps we do this partly because it is impossible to understand everything. It is in this spirit that we analyze probabilistic causality in terms of physical probability. If the world is deterministic the physical probabilities are only "pseudoprobabilities", a term that is appropriate by analogy with "pseudorandom numbers". (Compare Good 1972b, p. 128; 1983a.) But we drop the prefix "pseudo" partly for the sake of linguistic simplicity, and partly because the world might well be indeterministic, as most physicists believe. When assigning credit and blame in ordinary affairs, in scientific endeavors, and in legal matters, we need to use judgment when deciding how much detail to take into account. That is why a judge so often decides that the jury should *not* be allowed to hear "the whole truth". If the whole truth were presented, each trial would last forever, and, at the end of eternity, an accused person (long since dead) would be acquitted by Madame de Staël's law. So, in the spirit of Bentham's "fictions" (Ogden 1932; Vaihinger 1911; Good 1969, 1985d), we shall regard causal tendency as a physical phenomenon, depending on physical probabilities although, in every application, human judgement and common sense are involved. Even if the world is deterministic it behaves as if it were *indeterministic relative to a given depth of analysis*. We use subjective probabilities that are our best estimates of logical probabilities and I shall regard these, in the spirit of statistical mechanics, as averages of propensities over all worlds consistent with the level of detail attained. Part of the justification of this point of view is that the world might well be indeterministic in which case we really are trying to estimate physical probabilities. Since the question of whether the world is deterministic is metaphysical, we may as well regard the logical probabilities as estimates of physical ones. In other words, from a pragmatic or "as if" point of view, credibilities are rough estimates of propensities.

People do not necessarily consciously estimate probabilities when they make decisions and when they apportion credit and blame. Even a wasp behaves as if it assigns blame for its discomfort when it stings you, but the wasp has not read a book on probability unless it was a sarcastic statistician or philosopher in a previous life, now punished by being a wasp. Just as people and wasps behave to some extent *as if* they relied on subjective probabilities and utilities, I think they also behave as if their estimates of causal tendencies are based on probability estimates.

Since the brain contains millions of neural networks it is tempting to speculate that it contains networks resembling some of the causal networks in the outside world. Presumably it will be a long time before this speculation will be properly investigated even in primitive animals although I understand that the nervous system of a nematode has been fairly completely mapped out in Cambridge by Sydney Brenner's group. (Compare Good 1965, pp. 54–55, and 67–69.[8]) Like any speculation about the nervous system, right or wrong, this one might be used for designing machines.

One reason for trying to make the concept of probabilistic causality quantitative is that, if the quantitative explication is satisfactory, then its qualitative implications will also be acceptable. Conversely, a good qualitative theory ought to be embeddable in a quantitative theory, though not necessarily a unique one. Some aspects of the qualitative explication of causality by Suppes (1970) are implied by my more quantitative discussion. The relationship between our approaches is discussed in my review of Suppes's book (Good 1972a) and I shall say no more about it in the present paper.

I had originally hoped to review all the literature on probabilistic causality, but I soon discovered that reviewing my own efforts would more than fill the time. In particular, I shall say very little about path analysis which was introduced by Sewall Wright (1918) in connection with population genetics and which is designed to measure causal influence. Many sociologists have become interested in the technique. (See Tate 1978, for a survey.) Path analysis requires the user to construct a causal net before doing the calculations. These calculations depend on correlations and partial correlations, or on regression theory. The linearity assumptions that are usually made are not required in my analysis.[9] I shall also not discuss the work of Norbert Wiener (1956, pp. 184–190) on degrees of causal influence between time series which also depends on a form of regression theory. (See also Granger and Hatanaka 1964; Granger's chapter in this book; and Deaton and Foutz 1980.) A connection of the present work with regression theory is shown in Example (vi) below.

Another related topic is what used to be called "the probability of causes" which refers to the use of Bayes's theorem for converting prior (initial) probabilities of hypotheses into posterior (final) probabilities. When there are just two hypotheses Bayes's theorem is closely related to the concept of weight of evidence which is discussed in Section 3.

It is essential to distinguish between the tendency of F to cause E and the degree to which F actually caused E, denoted by $\chi(E:F)$. (Phonetically Q and χ initialize the word "causality". The historical reason for preferring the symbols Q and χ to C is mentioned by Good 1961, p. 307.) The distinction is like that between attempted murder and actual murder.[10] Another clear example of the distinction between Q and χ is given by Good (1961, p. 318; or 1983b, p. 216) in which Watson tries unsuccessfully to prevent Moriarity from killing Holmes. A similar example, suggested by a colleague, is an unsuccessful operation for which the prior chance of success was high: the surgeon is not necessarily blameworthy for his failure although he was part of the causal net. Perhaps an earth tremor was largely responsible. Let us consider a further example:[11] A good golfer, 70 yards from the green, hits the ball (event F) with the intention that it will land on the green (event E). Assume that $P(E|F)$ is large. But the ball happens to hit a bird and drops to the ground where a chimpanzee, that has escaped from a language laboratory at the local zoo, then obligingly carries the ball to the green. Here $Q(E:F)$ is appreciable but $\chi(E:F)$ is negligible or zero because the chain of causation has been cut or nearly cut.

In Britain, if two ships collide, the Admiralty attempts to apportion the blame between the two captains. This is not the same as estimating the probability that one captain was the man or woman uniquely responsible, because both captains might be partly responsible, one more than the other. Perhaps the allocations should be in the same ratio as the two tendencies to cause the collision provided by the actions taken, or failures to take action, by the two captains. Hart and Honoré, in their very interesting book on causation in the law (1959, pp. 214–215) ask "Can a meaning be attached to 'degrees of causation'?" and they mention that apportionment of blame is required by most Canadian and American statutes. Perhaps the courts should try to estimate the ratio of the Q's more than that of the χ's. But the legal problem is complicated because the law is concerned with all four of χ, Q, the amount of damage done, and the prior expected amount of damage.

From a philosophical point of view, whether or not the application is legalistic, it is desirable to consider a quantitative or semi-quantitative explication of causal tendency because it is unsatisfactory to know that one event has some tendency to cause another event without knowing, for example, whether the tendency is large or negligible.

The philosophy of strict causality should be a *limiting* case of probabilistic causality in the sense that "certainty" is to be replaced by a probability very close to 1, such as $1 - 1/10_4$ where $10_0 = 1$, $10_{n+1} = 10\text{xp}(10_n)$, $b\text{xp}(a) = b^a$ $(n = 0, 1, 2, \ldots)$. With this convention, if F is a sufficient cause of E, we find that there is a virtually infinite weight of evidence against F if E does not occur. But if F is only a necessary cause, then the magnitude of the causal propensity can be small or large and our explication of Q is consistent with this requirement. Strict causality will not be discussed further in this paper.

We'll soon need a definition of "weight of evidence", but for the moment let's leave it informal because it's more vital to discuss the incompleteness of the notation $Q(E:F)$. To begin with it leaves out of account the physical state of affairs U (the state of the universe) just before F occurred. All the probabilities must be made conditional on U. They must also be conditional on all true laws of nature, say H. To see the need to allow for U and H, consider the following example that was used by Salmon (1980) in an honest attempt to refute the explicatum of $Q(E:F)$ that I had proposed. He said that a "falling barometric reading has no tendency at all to cause a storm, though the barometric reading is highly relevant statistically to the onset of stormy weather". Likewise Russell (1948, p. 333) cites a colleague who said in despair "The barometer has ceased to have any effect on the weather". This example refutes any explicatum that is *only* a measure of statistical association between F and E, but the explicatum I advocate is more than that because the probabilities are conditional on U and H. This makes $Q(E:F) = 0$, as it should be, in the example of the barometer.[12] So please bear in mind that a fuller notation is $Q(E:F|U\&H)$ where the vertical stroke is now a standard notation and is pronouned "given" or "conditional upon".[13] An even more complete notation for $Q(E:F)$ will be mentioned in Section 3.

Salmon had some valid criticisms of my work on χ, to be discussed further in this volume.

2. WEIGHT OF EVIDENCE

It is necessary next to define or explicate "weight of evidence". Let G denote the proposition that an accused person is guilty, and let C denote the evidence presented in court (G for Guilty, C for Court). Some of this evidence might support G and some might undermine it, and these can be regarded as positive and negative pieces of evidence. I don't interpret the expression "weight of evidence" as proportional to the weight of the paper that the evidence is written upon. Such a definition would not be worth the paper on which *it* is written. Even J. M. Keynes (1921, pp. 71–78, 312), who toyed with essentially this uninteresting interpretation, though he usually called it "the weight of argument", said "I remain uncertain as to how much importance to attach to it", that it "cannot be explained in terms of probability", "I do not feel sure that the theory of 'evidential weight' [in his sense] has much practical significance", "The weight [in his sense], to speak metaphorically, measures the *sum* of the favourable and unfavourable evidence, the probability measures the *difference*," and "The magnitude of the probability of an argument [? hypothesis] ... depends upon a balance between ... the favourable and the unfavourable evidence." The usual interpretation of weight of evidence, in ordinary English, has indeed the sense of the balance of the evidence and that is my interpretation.

The weight of evidence in favor of G (or the support for G) *provided by C*, or the corroboration of G by C, denoted by $W(G:C)$, should be some function f of $P(C|G)$ and $P(C|\bar{G})$ alone, where $\bar{G}$ means that the accused is innocent (not guilty). There will also be background information B which can be taken into account in the notation by writing $W[(G:C)|B]$ or, more briefly, $W(G:C|B)$ if we regard the colon as more "binding" than the vertical stroke. But I am omitting B to simplify the notation. (An example of background information would be information about how often people are "framed".)

Our first assumption about W is then

$$W(G:C) = f[P(C|G), P(C|\bar{G})]. \quad (1)$$

We have assumed that W does not depend also on the prior probability $P(G)$ once $P(C|G)$ and $P(C|\bar{G})$ are taken into account. A justification for this assumption is that we are discussing the contribution of C alone to the posterior probability of G (given B). But when judging this posterior probability, if we are rational, we must of course take into account both the prior probability of G and the weight of evidence provided by C. That is, for some other function ϕ, we would like to have

$$P(G|C) = \phi[W(G:C), P(G)]. \quad (2)$$

It will help the judgement if the functions f and ϕ are simple, but we do not for the moment make this assumption.

From the two assumptions (1) and (2) it can be proved mathematically (see Good 1968, Appendix F; and especially 1984) that $W(G:C)$ must be a function of $P(C|G)/P(C|\bar{G})$. If we are mainly concerned with ordering weights of evidence we may as well take the logarithm

$$W(G:C) = \log[P(C|G)/P(C|\bar{G})], \quad (3)$$

because this implies the convenient additive property

$$W[G:(A \& C)] = W(G:C) + W(G:A|C). \quad (4)$$

In particular, when A and C are statistically independent given G and also given $\bar{G}$, we have

$$W[G:(A \& C)] = W(G:C) + W(G:A). \quad (5)$$

The additive property, in either form, is desirable when you think of the Goddess of Justice, called Themis by the ancient Greeks, with two scales, for adding up the balance of the evidence which, as already mentioned, is another name for the weight of evidence. For example, smoking and living in a highly urban area have been found to provide roughly additive weights of evidence in favor of lung cancer. (See Buck and Wicken 1967, and Good 1967.) The additive property is also desirable if we want to use weight of evidence as a quasi-utility or epistemic utility whose expectation can reasonably be maximized in the design of experiments (Good 1979b). The additive property could be regarded as a subsidiary desideratum. Therefore, of all explicata of weight of evidence, (3) is the best one if my desiderata are accepted. The formulae $P(G|C) - P(G)$ and $P(G|C)/P(G)$, for weight of evidence, which some philosophers have been using, even as late as 1985, both violate condition (1). For some arguments against them see Good (1985a, f). A further disadvantage of $P(G|C) - P(G)$ is that when $P(G|C)$ and $P(G)$ are both small, the difference is also small even if the ratio is large. If two independent pieces of evidence, C_1 and C_2, would each by itself multiply the odds of G by a million, then, as is easily proved, when combined they multiply the odds by a trillion. (Odds means $p/(1 - p)$ where p is a probability.) Are we to regard C_1 as of small weight if it was presented first and raised the odds from 10^{-9} to 10^{-3}? It seems to me that C_1 and C_2 have equal weights, whichever of the two is stated first. So the difference $P(G|C) - P(G)$ seems to me to be clearly a bad measure for the weight of evidence of C, and highly inconvenient to use. Similarly the ratio is also seen to be a bad measure if $P(G|C)$ and $P(G)$ are both close to 1 while $P(\bar{G}|C)/P(\bar{G})$ is large. Thus $P(G|C) - P(G)$ and $P(G|C)/P(G)$ can both be unreasonably insensitive to what are intuitively large weights of evidence, at least in my opinion.

If a piece of evidence is independent of all other evidence (given G, and also given $\bar{G}$) then its weight does not depend on when this evidence is presented when W is the explicatum.

Expected weight of evidence has been given other names, such as cross entropy, relative entropy, and the Kullback expression, and plays a prominent role in statistical theory.

For a review paper on weight of evidence see Good (1985a).

3. THE FORMULA FOR CAUSAL TENDENCY, Q

When the mentioned interpretation for weight of evidence is used the formula for $Q(E:F)$, previously expressed in words, becomes $W(\bar{F}:\bar{E})$, that is (but don't be scared by the notation),

$$Q(E:F) = W(\bar{F}:\bar{E}) = W(\bar{F}:\bar{E}|U \,\&\, H) = \log\left[\frac{P(\bar{E}|\bar{F} \,\&\, U \,\&\, H)}{P(\bar{E}|F \,\&\, U \,\&\, H)}\right]. \tag{6}$$

One could use the right side of (6) as the definition but the intuitive appeal is greater, and the concepts are simplified, when the interpretation in terms of weight of evidence is held in mind.

There is often some question about the meaning of the negation of a proposition. Consider, for example, the proposition or event that a person smokes an average of forty cigarettes per day. We might want to contrast this with his being a moderate smoker or a non-smoker. Similarly, in the example of the firing squad, the negation of dropping dead might be living for a minute or more before dying, or it might be a disjunction of many possibilities. When we say that F tends to cause E we must have in mind some concepts of the negations of F and of E. The concepts might be vague, but for a formal expression we must use symbols for these negations. In a rigorous analysis, and at the risk of scaring the reader away (indeed, at both the PSA meeting and at the Irvine meeting, one member of the audience walked out at this point), one should mention a notation (which contains no mathematics) such as

$$Q[(E/E'):(F/F'_D)\,|\,U \;\&\; H \;\&\; (E \vee E') \;\&\; (F \vee F'_D)], \qquad (7)$$

or, in a more friendly notation,

$$Q[(E/E'):(F/F'_D)\,|\,U,\, H,\, E \vee E',\, F \vee F'_D], \qquad (7\tfrac{1}{2})$$

the causal tendency to produce E rather than E' provided by F rather than F'_D given the state of the universe just before F occurred, the true laws of nature, and the assumption that either E or E' will occur and that either F or F'_D had to occur. Here F'_D denotes a disjunction of possibilities over which there is a conditional probability distribution, D, conditional on F not occurring. For the sake of simplicity of notation I call E' and F'_D the negations of E and F and denote them by $\bar{E}$ and $\bar{F}$. Note the distinction between the oblique stroke ("rather than"), the vertical stroke ("given") and the colon ("provided by" or "contributed by"). These distinctions are useful in other contexts where weight of evidence or amount of information is the main topic instead of causality.

One must remind oneself from time to time that $Q(E:F)$ depends on what we choose to regard as the alternatives to E and F. The *meaning* of Q depends on this choice. The throwing of a loaded die (event F) is a (probabilistic) cause of its coming up 4 (event E), even if the probability of a 4 is only 1/100, provided that we regard the alternative to F as not throwing the die. But if the alternative is throwing a fair die, then F was a "negative cause" of E. It can hardly be emphasized too much that $Q(E:F\,|\,U \;\&\; H)$ is an incomplete notation unless $\bar{E}$ and F'_D are implicitly specified. Similar remarks apply to χ.

There is often also a problem about the meaning of an event. An assassin is not responsible for every detail of the victim's death. The victim's death is a disjunction of a very large number of detailed events. If we define the event E in too much detail we'll find that the assassin's behavior had negligible tendency to cause E if Q is defined by equation (6). The assassin is not tried in court for shooting the victim in the third rib from the top, four inches from the sternum; he is tried for murder. The same kind of reliance on common sense is required in probability theory when we talk about repeating an experiment "under essentially the same conditions". Unfortunately there are many circumstances where it is difficult to decide whether two events are essentially the same. This introduces a further subjectivistic aspect into the philosophy of causality which I believe is inevitable. This point is emphasized in the appendix in relation to degrees of causation, χ, instead of causal tendencies.

One *inadequate* attempt to avoid the difficulty mentioned here would be to define the tendency of F to cause E as the weight of evidence in favor of F provided by E. Then it wouldn't matter if E were defined in intricate detail. A decisive objection to this definition can be seen by an example: it would force us to say crossing a road has a large tendency to cause an accident, whereas it is clear to common sense that the tendency is small in most circumstances.

I think formula (7) by itself encapsulates about 20% of what I mean by the philosophy of causality and equation (6) another 25%. It is easy to overlook the need to make Q depend on all the arguments mentioned in formula (7), but I think this dependence is convincing once it is pointed out. The need to introduce alternatives to E and F is somewhat analogous to the requirement to consider non-null hypotheses when applying a significance test of a null hypothesis, a clear requirement both in the Bayesian and the Neyman–Pearsonian approaches.[14]

4. THE DESIDERATA

It would take too much time to give the full argument for formula (6) so let's be satisfied with an outline. For the sake of simplicity I am going to make somewhat stronger assumptions than in my previous work (Good 1961 and 1980e; or 1983b, pp. 197–218). But, as in that work, the argument depends on an analogy between causal networks and electrical networks, though the analogy is far from perfect.

When electrical resistances are put in series, the resultant resistance is equal to the sum of the individual resistances; whereas, when they are in parallel, the reciprocal of the resultant is equal to the sum of he individual reciprocals (Roy. Corps Sig. 1947, pp. 124–125). Thus, if we define the *strength* of an electrical network as the reciprocal of its resistance, we have an additive property of strength for links in parallel. Can we reasonably assume something similar for causal networks?

A causal network or net will be defined as a collection of events of interest in some causal context, where each event is possible but may or may not have occurred (or will have occurred). The network is supposed to start at an event F and end at an event E. To have $\chi > 0$, it will be necessary that the causal net should *contain* a connected set of events, from F to E, all of which occurred. More is said about χ in the appendix. Note that Q (but not χ) can be positive even if none if the events in the network occur.

A causal network will be assumed to have a non-negative causal strength $S = S(E:F)$ and a non-negative resistance $R = R(E:F)$, where R and S are related by a monotonic decreasing functional relationship but the relationship will not turn out to be $S = 1/R$. The resistance of a chain of links (links in series) is assumed to be equal to the sum of the resistances of the links (see Figure 1).[15] If the resistance of one of the links is infinite then the chain is cut, and the resistance of

Fig. 1. A chain consisting of three links. A chain is no stronger than its weakest link. The resistance of a chain is the sum of the resistance of its links.

the whole chain is infinite and its causal strength is zero.[16] The strength of a chain consisting of a single link $F \rightarrow E$ is assumed to be equal to $Q(E:F)$. We may think of $\chi(E:F)$ as the strength of the network when all details are filled in.

Consider two chains "in parallel" as in the "firing squad" illustrated in Figure 2. Assume that G_1 and G_2 are independent given F and are also independent given $\bar{F}$. (In other words there is a "conjunctive fork" at F in the terminology of Reichenbach 1956, p. 159.) Assume further that $P(E|G_1) = P(E|G_2) = 1$, that is, that both marksmen are crack shots. When these conditions are satisfied we call G_1 and G_2 "independent causes" of E (as defined more generally in Good 1961, p. 45; or 1983, p. 211) and we assume that the strength of the net is the sum of the strengths of the two chains and is also equal to $Q(E:F)$. This is the strongest assumption that I make and therefore the weakest link in my argument. Without this relationship between S and Q I have been unable to obtain any sensible formula for Q, but for general networks S is by no means the same as Q. For general networks S and

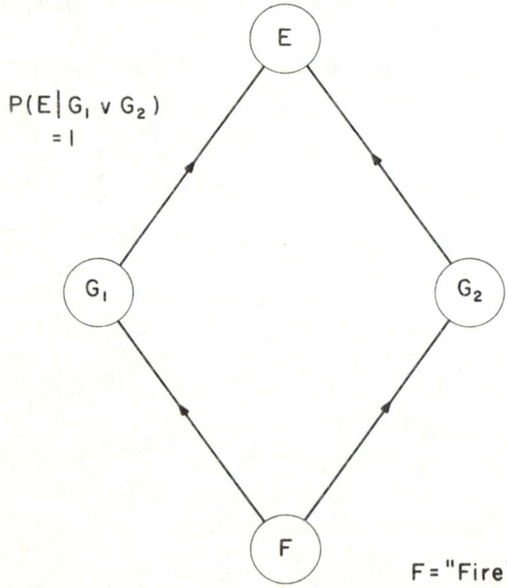

Fig. 2. A firing squad, illustrating chains in parallel.

χ depend on events that might occur later than F and earlier than E, but Q does not depend explicitly on these intermediate events.[17] For the firing squad, however, it seems reasonable enough to collapse F, G_1 and G_2 into a single event because we could imagine the captain himself as doing the shooting, with one gun in each hand.

An example similar to the firing squad is illustrated in Figure 3.

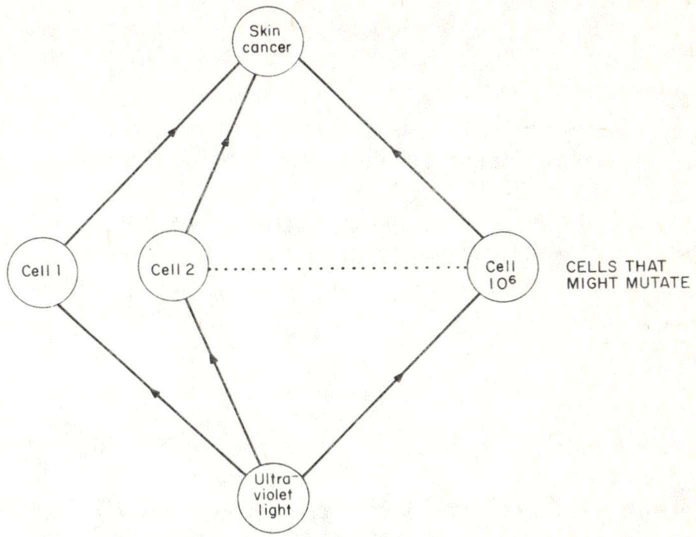

Fig. 3. Another example of chains in parallel. The strength of the network is the sum of the strengths of the million chains.

Assume further that if $F \to G \to E$ is a causal chain, and if $(F \,\&\, E)$ implies G, then the chain is of the same strength as the collapsed chain $F \to E$.

Consider now a radioactive particle in a certain state, called white, which has probability $e^{-\alpha t}$ of remaining in that state for time t or longer, otherwise it becomes permanently black. Let F be the event that it starts in the white state and E the event that after time T it is also in the white state. We can think of the chain of events that specify the "color" of the particle at times $0, \delta t, 2\delta t, \ldots, T$ where δt is small. This causal chain has a strength that is assumed to tend to a limit S

when $\delta t \to 0$ where the limit does not depend on the unit used to measure time. This assumption provides a constraint on the explication of the strength of a causal chain.

Somewhat surprisingly, with only the addition of minor assumptions, those made so far imply the following result (Good 1961, 1980e):

THEOREM. (i) *$Q(E:F)$ is given by equation* (6); (ii) *R and S are related by the equations*

$$e^{-R} + e^{-S} = 1, \qquad R = -\log(1 - e^{-S}),$$
$$S = -\log(1 - e^{-R}).^{18} \tag{8}$$

Based on this theorem a sort of causal calculus can be developed, but further assumptions are required for defining the strength of more general causal networks (Good 1961), and hence for defining χ. Indeed I am uneasy about the part of the definition of χ for entirely general networks that I gave in Good (1961, pp. 49 and 88; or 1983b, pp. 214–215). In most of what follows, other than in the appendix, I discuss only causal tendency Q.

5. EXAMPLES

(i) *A sweepstake.* Let F be the act of buying a sweepstake ticket and E the event of winning the first prize. Your failure to win is only slight evidence that you did not buy the ticket. Thus $Q(E:F)$ is small in this example (even if you win). But if you do win, the purchase of the ticket is a large part of the *explanation*. Causal propensity is not the same as what I call "explicativity" (Good 1977, p. 317).[19] One distinction between causal propensity Q and explicativity is that a causal propensity cannot be far-fetched. This is because $Q(E:F)$, unlike explicativity, *does not depend on the prior probability of F* (although it depends on the distribution D mentioned earlier). For example, if a brick hits a window, and we don't know why, it is usually a better explanation to say that a naughty boy threw the brick than that the Mother Superior threw it: for one thing she is known to be a bad shot. But if in fact she did throw the brick, and accurately, then her surprising performance was a cause of the lamentable incident. It is the *suggestion* of an improbable cause that can be far-fetched, not the cause itself. A suggestion of a probable cause can be a good explanation.

(ii) *A two-state Markov process.* We can generalize the example of the radioactive particle by allowing return from the black to the white state, with parameter β. We previously had $\beta = 0$. Then it can be shown that

$$Q(E:F) = \log[(\alpha + \beta e^{-(\alpha+\beta)T})/(\alpha - \alpha e^{-(\alpha+\beta)T})]. \tag{9}$$

(This is log coth(αT) if $\alpha = \beta$.) As T increases from 0 to ∞, Q decreases from ∞ to 0.

(iii) *Components of causal influence.* The following identity follows from (6):

$$Q[E:(F \,\&\, G)/(\bar{F} \,\&\, \bar{G})] = Q(E:G|\bar{F}) + Q(E:F|G). \tag{10}$$

For example, the tendency for lung cancer to be caused by smoking and living in a smoky environment, as against not smoking and living in a clear environment, is equal to the tendency through living in a smoky environment, given no smoking, plus the tendency through smoking, given that the environment is smoky.

A simpler decomposition than (10) is satisfied by the expression

$$K(E:F) = \log\frac{P(\bar{E})P(F)}{P(\bar{E} \,\&\, F)} = \log\frac{P(F)}{P(F|\bar{E})},$$

which is related to Q in essentially the same way that information I is related to evidence, W, where

$$I(A:B) = \log\frac{P(A \,\&\, B)}{P(A)P(B)}.$$

Thus $K(E:F)$ is defined as the mutual information between F and not E (given U and H). We may call K "intrinsic tendency to cause" because it resembles Q but avoids reference to the negation of F. It has the decomposition

$$K[E:(F \,\&\, G)] = K(E:F) + K(E:G|F). \tag{11}$$

K might be more useful than Q when we don't want to specify the alternatives to F. There is an analogy here with the use of Fisherian significance tests when the alternatives to the null hypothesis are extremely vague so that full-dress Bayesian and Neyman–Pearsonian methods are hard to apply. This somewhat weak analogy between the philosophy of causal propensity and that of testing hypotheses is interesting but I'll say no more about K in this paper.

(iv) *A 2-by-2 population contingency table* (Good 1981). There is a large literature on measures of association for the modest 2-by-2 contingency table, even when the sample is so large that we can call the table a "population contingency table". For example, Yule (1903, 1911), Goodman and Kruskal (1954, 1959), Sheps (1958, 1959), Greenberg (1969), and Fleiss (1973). In particular, Yule (1903) and Yule and Kendall, presumably in all editions of their well-known textbook (e.g. 1950, pp. 36–38), emphasized that spurious associations can appear if two populations are amalgamated, a point made later by Cohen and Nagel (1934, p. 449) and by Simpson (1951). See also Good and Mittal (1985) where conditions for avoiding this "amalgamation paradox" are discussed in detail.

With this warning made, consider a 2-by-2 contingency table based on a random sample from a homogeneous population,[20] and having

Maltreatment	Disease or Death: D	$\bar{D}$	
M	a=400	b=199600	a+b =200000
$\bar{M}$	c=160	d=799840	c+d =800000
	a+c =560	b+d =999440	a+b+c+d =1000000

Fig. 4. A two-by-two table.

cell entries *a*, *b*, *c*, *d* in typewriter order. Assume that the sample is large enough so that *a*, *b*, *c*, and *d* can be taken as proportional to the probabilities or fractions in the population. See Figure 4, where some imaginary numerical values are assumed. The first row corresponds to a maltreatment or ill-treatment *M* such as smoking or radiation, the second row to $\bar{M}$. The first column corresponds to disease or death *D*, and the second column to $\bar{D}$. (This meaning for *D* is of course different from the meaning in formula (7).)

Of the various measures of association for contingency tables, some are symmetrical between rows and columns. Edwards (1963) showed that a measure of association must be a function of the odds ratio, *ad*/(*bc*), under assumptions implying that the measure is unchanged when the rows are interchanged with the columns. This is often a reasonable property but not when we are concerned with causality or weight of evidence.

The degree to which the ill-treatment tends to make people ill is, in decimal units,

$$Q(D:M) = \log_{10}\left(\frac{d}{c+d}\Big/\frac{b}{a+b}\right) = 0.00078 \text{ decimal units}$$
$$= 0.78 \text{ millicausats}, \quad (12)$$

to coin a term. I noticed for the first time in 1970 (Good 1971, p. 114) that this is the logarithm of one of the measures of association that have been suggested: *the survival ratio* (Greenberg 1969; Sheps, 1958, 1959). (A more general name would be *benefit ratio*.) Different measures measure different things, a point emphasized by Goodman and Kruskal (1954). For prognosis, the weight of evidence in favor of *D* provided by *M* is

$$\begin{aligned} W(D:M) &= \log\frac{P(M|D)}{P(M|\bar{D})} = \log\frac{O(D|M)}{O(D)} \\ &= \log\left(\frac{a}{a+c}\Big/\frac{b}{b+d}\right) \\ &= 0.55 \text{ decimal units ("bans")} \\ &= 55 \text{ centibans} \qquad (13) \end{aligned}$$

the antilogarithm or Bayes factor being 3.6. (The name "ban" is due to

Turing. See Good 1979b.) For detective work, the weight of evidence in favor of M provided by D is

$$W(M:D) = \log\frac{P(D|M)}{P(D|\bar{M})} = \log\frac{O(D|M)}{O(M)}$$

$$= \log\left(\frac{a}{a+b}\Big/\frac{c}{c+d}\right)$$

$$= 1 \text{ ban} = 100 \text{ centibans}, \tag{14}$$

corresponding to a Bayes factor of 10.

The causal tendency is very small because, in this example, D is fairly rare. But consider the following example.

(v) *Russian Roulette.* (See Figure 5, where of course R has a new meaning.)

Here $P(E|R) = 1/6$, $P(E|\bar{R}) = 0$, and

$$Q(E:R) = W(\bar{R}:\bar{E}) = \log\frac{R(\bar{E}|\bar{R})}{P(\bar{E}|R)} = \log\frac{6}{5}$$

$$= 8 \text{ centicausats} \tag{15}$$

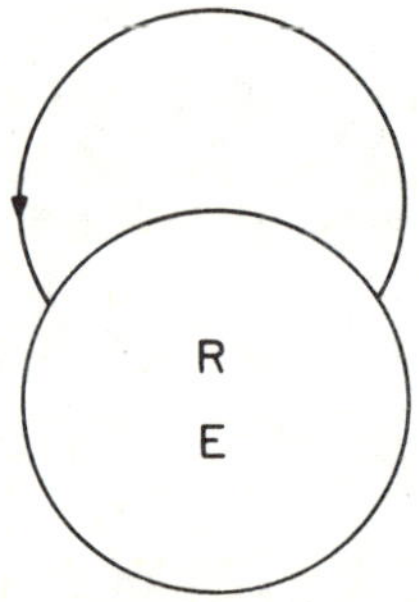

Fig. 5. Russian roulette.

or 8 centibans. (It is a weight of evidence so the units can be given this name but it might be misleading to do so because it is not the weight of evidence concerning E provided by R.)

In Figure 6 a commissar says "Fire" to three marksmen, each of whom plays non-auto-Russian Roulette. The judge fixes the number of marksmen depending on his opinion of the severity of the crime.

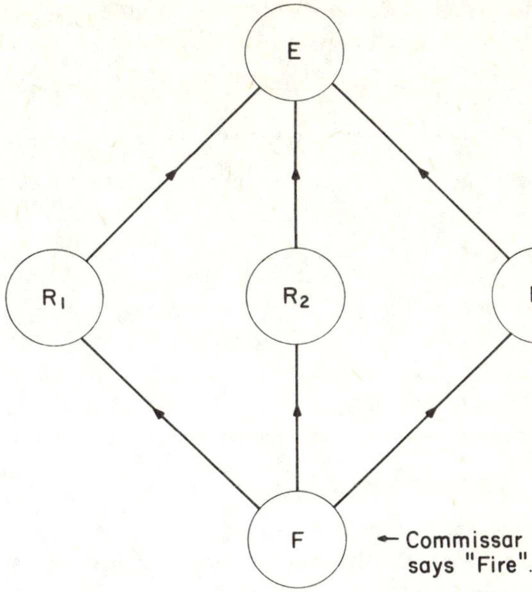

Fig. 6. Non-auto Russian roulette. The judge determines the number of marksmen.

Assume here that the marksmen fire if and only if the commissar tells them to, an assumption not made before for the firing squad. Here $Q(E:F) = 24$ centibans (8 cb for each chain), the antilog being 1.7. The weight of evidence in favor of the commissar having said "Fire", if E occurs, is

$$W(F:E) = \log[P(E|F)/P(E|\bar{F})]$$
$$= \log\{[1 - (5/6)^3]/0\} = \infty, \quad (16)$$

which means that the commissar certainly said "Fire".

(vi) *Linear regression* (Good 1971, pp. 114–115; or better, 1980c). Suppose $\mathbf{y}$ has the multivariate normal distribution of mean $B\mathbf{x}$ and covariance matrix C. (The reader might prefer to think first about the univariate model $\tilde{y} = \beta x + \tilde{\epsilon}$, where $\tilde{\epsilon}$ has a Gaussian distribution with mean 0 and variance σ^2. The tilde means we are dealing with a random variable, but I shall not always use the tilde.) Suppose the "treatment" $\mathbf{x}$ is changed to $\mathbf{x} + \boldsymbol{\xi}$ and is applied to essentially the same physical system. Since, for each $\mathbf{x}$, $\tilde{\mathbf{y}}$ is a continuous random variable, a definition of the causal tendency is required. The most natural definition is

$$Q(\tilde{\mathbf{y}}:\Delta\mathbf{x} = \boldsymbol{\xi}) = \mathscr{E}_{\boldsymbol{\eta},\boldsymbol{\eta}'}\, Q[(\tilde{\mathbf{y}} = B\mathbf{x} + B\boldsymbol{\xi} + \boldsymbol{\eta})/ (\tilde{\mathbf{y}} = B\mathbf{x} + \boldsymbol{\eta}'):\Delta\mathbf{x} = \boldsymbol{\xi}]$$

(in which the oblique stroke here means "as compared with")

$$= \tfrac{1}{2}\boldsymbol{\xi}'B'C^{-1}B\boldsymbol{\xi},$$

or $\beta^2\xi^2/(2\sigma^2)$ for the case of simple regression. The causal tendency is invariant under a linear transformation of the model and, in the multivariate case, is additive for two independent causes, that is, if the matrices B and C are diagonal. Also the causal tendency is unchanged if $\boldsymbol{\xi}$ or ξ is multiplied by minus 1.

If (x, y) has a bivariate normal distribution with correlation ρ, then

$$Q(y:\Delta x = \xi) = \rho^2\xi^2/[2(1-\rho^2)\sigma_x{}^2]. \tag{17}$$

By virtue of this connection between Q and linear regression, there is a prospect of relating the present theory to path analysis. There is also a strong analogy (Good 1980d) between path analysis and Feynman's "path integral" approach to quantum mechanics (Feynman 1948; Feynman and Hibbs 1965). The addition and product rules in path analysis and in Feynman's theory are also somewhat analogous to those in the present theory. So perhaps all three theories might be integrated, but this is speculative.

(vii) *Prospective versus retrospective studies.* I roughly quote here from my article on statistical fallacies (Good 1979a, p. 348).

We are often interested in the values of the two conditional probabilities, $P(E|F)$ and $P(E|\bar{F})$, without wishing to assign a quantitative measure to the degree of causation. These probabilities cannot be inferred solely from the values of $P(F|E)$ and $P(F|\bar{E})$. Therefore a retrospective study cannot tell us everything that a prospective one can (nor conversely, but that is less important). As Bortkiewicz (1911) said

> We might sample a group of idiots and find that their grandparents were mostly normal. It would certainly be absurd to infer from this that most of the grandchildren of normal people are idiotic. (English translation by W. H. Kruskal.)

But retrospective studies do not need to be used idiotically. A retrospective study can have more value per unit effort than a prospective study, especially of course when a prospective study is impracticable. We have

$$P(E|F)/P(E) = P(F|E)/P(F), \tag{18}$$

so if we discover by a retrospective study that $P(F|E) > P(F)$ then we can deduce that $P(E|F) > P(E)$, and therefore that F is a probabilistic cause of E, but not necessarily a strong one.

A probabilistic cause need not be a strong one in order to be important. Joshua Lederberg has mentioned in correspondence the following striking example of work by Herbst and others. They found that many mothers of daughters with vaginal cancer (event E) had been treated with DES (diethylstilbestrol) during pregnancy (event F). The word "many" here must be taken to imply that $P(F|E)/P(F)$ is much larger than 1, so that $P(E|F)/P(E)$ is also much larger than 1. Although this does not show that $P(E|F)$ is close to 1, it does show (if the observed facts are clear and without measurement artifact) that the administration of DES during pregnancy can cause much damage. Crossing a road can also cause much damage but we are not prepared to give up crossing roads: we just cross carefully.

These qualitative remarks could be made more quantitative by using the definition of causal tendency.

(viii) *Judgement of causal tendency.* It might be possible to cultivate a direct quantitative or semi-quantitative judgement of causal tendencies. In some sense we must already use such judgements implicitly. If some of these judgements can be made explicit they could be used to improve our judgements or probabilities just as other functions of probabilities, and utilities, can be used. Compare the discussion of the input to *de luxe* black boxes in Good (1983b, p. 76).

6. EXTENSION TO MORE GENERAL NETWORKS

In my 1961 paper I suggested a method for defining the strength of an arbitrary causal net. I shall not repeat the details here because they are somewhat intricate and unconvincing. This extension was necessary for defining $\chi(E:F)$ as the limit of the strength of the net joining F to E when the events in the net are made smaller and smaller. An important amendment is given in our appendix, and there is further clarification in my reply to Salmon's paper in this volume.

7. SUMMARY

My aim has been the quantitative explication of the notion of causal tendency in terms of probability, and to give examples of the preferred explicatum.

APPENDIX. DEGREE OF CAUSATION, $\chi(E:F)$

I have relegated the discussion of degree of causation, χ, to this appendix because my main topic has been Q and there are still unsolved problems concerning χ. The explicatum for Q does not depend on that for χ except perhaps for the connection to "strength" in relation to Figure 2.

When attempting to define $\chi(E:F)$ where $\chi > 0$ we must think of a causal net from F to E that contains at least the relevant set of events all of which actually occurred, as mentioned in the main text. The example of the assassin shows that we must also allow for various events that did *not* occur. This point is made extremely clear by means of the following example.

Consider a pinball table in which the ball exits through a particular hole (event E), so that playing the ball (event F) leads pretty well inevitably to E. The degree to which F caused E is high, but the degree to which F caused the precise path is low. A similar example is provided by the boiling of a kettle of water by applying heat. On the assumption of an indeterministic world, the degree of causation is negligible for the specific manner in which the water boiled (with a complete description of the motions of all the molecules). Yet the degree of causation of the *boiling* was great. In both this example, and for the pinball table, there seems at first sight to be something to be said for the remark by Salmon (1980) that the degree to which F actually caused E should depend partly on $P(E|F)$. But it clearly hardly depends explicitly on $P(E|F)$ in the example concerning the lucky golfer and the obliging chimpanzee, and I don't believe it ever should depend explicitly on $P(E|F)$. We appear for the moment to have a new riddle of causation.

Part of the resolution of this riddle is that we humans lump together a vast collection of essentially similar processes when considering whether one event was a cause of another one. We can define $\chi(E:F)$ as the strength of the causal net from F to E when all details are filled in, but we have to use common sense when defining the causal net. The causal net must contain all subnets, such as paths, that in some sense are *similar* to the relevant subnet of events that actually occurred (Good 1985e). The need to use common sense is clear in some legal problems: see, for example, Hart and Honoré (1959, p. 67). In the golf example, $\chi(E:F)$ would be increased, while still remaining negligible, if the bird were one of a large flock of birds, and if several chimps had escaped, as a group, from the language laboratory where they had learnt to play golf. In the pinball-table, assassin, and boiling-water examples, $\chi(E:F)$ is large when we allow for all possible paths of the ball or all paths of the bullet resembling the path actually taken, or all paths in phase space, but χ would not be large if we allowed only for the paths that occurred.

In Good (1961, pp. 45–50; or 1983b, pp. 210–215), where the very difficult problem of giving a complete definition of χ is attacked, I overlooked the need to allow for events that did not occur. (See Note 13.) The causal nets there need to be enlarged in the manner we have just discussed. I said there that "whether degrees of causality exist is a matter of physics", but in view of the need to allow for "essentially similar processes", it seems that χ must depend also on human interests. On the other hand Q is close to having a purely physical meaning.

The ideas in this appendix are developed further in my reply to Salmon's paper in this work.

Departments of Statistics and Philosophy
Virginia Polytechnic Institute and State University
Blacksburg, VA.

NOTES

[1] This work was supported in part by an N.I.H. grant number 18770. A preliminary version of the paper was presented, by a 27-minute audiotape, at a session on Probability of Causation organized by Peter G. Groër for the Annual Meeting of the Society for Risk Analysis, 1984 September 31 to October 3, Knoxville, Tennessee. The next version was Good (1985b) which this chapter resembles, but there are numerous changes. One minor change is the replacement of the expression "causal propensity" by "causal tendency" because "propensity" has recently been frequently used by philosophers to mean physical or material probability and will occasionally be used in this sense in this paper (not necessarily interpreted as a long-run frequency).

[2] An argument is given in Good (1980c) that if F is allowed to occur later than E, the explicatum of Q does not absolutely rule out the possibility of precognition. For speculative discussions of "reversed causality" and "backward time" see Good (1962, pp. 152, 155, 179, 315, 326, and 328).

[3] For my views on kinds of probability see Good (1959, or 1983b, indexes). Whether physical probability can be in effect defined in terms of subjective probability, *à la* de Finetti, will not be discussed.

[4] For example, Savage (1954) and Cox (1961).

[5] For example, Good (1966).

[6] For example, Good (1984).

[7] For example, Good (1977).

[8] The formula in Good (1965, p. 67, line 7 up) is wrong. The correct formula is in Good (1961, p. 45) or Good (1983b, p. 211).

[9] An argument is given by Good (1978) that suggests that path analysis may be approximately applicable when the linearity assumptions are replaced by monotonicity.

[10] These are perhaps equally culpable but not in the eyes of the law. If a skillful surgeon saves the life of a criminal's victim he might also save the life of the criminal. It seems odd that the punishment of the criminal should depend on the skill of the surgeon. In such cases the criminal should in my opinion pay tenfold for the operation if it is successful.

[11] A similar golfing example was given by Suppes (1970, p. 41), based on an example of Deborah Rosen.

[12] Salmon said in a letter of 1980 July 5 that he had overlooked that the probabilities were intended to be conditional on U and H.

[13] Salmon raised a second (intuitive) objection to the interpretation of χ given in my 1961 paper. In my response, Good (1980a), I said that I thought he was unconsciously confusing Q with χ but I think I was unfair because I now suspect that we all confuse them under certain circumstances. His position, and my further response are contained in the present volume. Furthermore, in the discussion in Chicago of the version of the present paper that appears in *PSA* 84, Salmon correctly criticized the definition of a causal net that I had given in previous work. His objection was that I had said that the net consisted only of events that occurred. I had intended this remark to apply to χ, not to Q, but had not made this clear. For Q, none of the events need to "obtain". Even for χ I had made an error, as pointed out in the Appendix, the substance of which was circulated just before the Chicago conference. I have no reason so far to be worried about the explicatum for Q.

[14] Fisher did not make this requirement explicit in his writings but he must have been almost fully aware of it. This is clear if only from the fact that sometimes he would use a single-tail probability and sometimes a double tail.

[15] In my 1961 paper I made weaker assumptions than additivity.

[16] For further discussion of cutting and mending causal chains see Humphreys (1980) and Good (1980b). Hart and Honoré (1959, p. 5) use the expression to "break the chain of causation" which is apparently standard usage in legal contexts.

[17] When estimating the probabilities that define Q, in any concrete situation, it may be helpful to imagine possible networks connecting F to E, so there can be an implicit dependence on intermediate events. But a general formula for Q in terms of probabilities does not need to mention these intermediate events explicitly, and formula (6) does not do so.

[18] One of the simplest examples to illustrate formula (8) is that the strength of a chain consisting of links of strengths S_i ($i = 1, 2, \ldots$) is $-\log \Pi[1 - \exp(-S_i)]$. Consider, for example, what happens if one of the S_i's is zero or infinity.

[19] The informal definition of "explicativity" given by Good (1977) was "the extent to which one proposition or event F explains why another one E should be believed, when some of the evidence for believing E might be ignored". See also Good and McMichael (1984) and Good (1985c).

[20] By a homogeneous population I mean one in which the relevant physical probabilities are the same for each individual. When a Bayesian regards the individuals of a population as permutable or exchangeable, then he can be regarded as believing that the population is homogeneous. Compare Lindley and Novick (1981). Good and Mittal (1985) define homogeneity *with respect to specific parameters*, thus giving the concept of homogeneity a somewhat more general meaning.

REFERENCES AND BIBLIOGRAPHY

(This is not intended by any means to be a comprehensive bibliography, but it covers nearly everything I have written on probabilistic causality.)

Bartlett, John (1980) *Bartlett's Familiar Quotations*, 15th edn. (E. M. Beck *et al.*, eds.; Boston: Little, Brown & Company).

Bortkiewicz, Ladislaus von (1911) 'Discussion in a conference', *Schriften des Verein(s) für Sozialpolitik* **138**, 175–176.

Buck, S. F. and Wicken, A. J. (1967) 'Models for use in investigating the risk of mortality from lung cancer and bronchitis', *J. Roy. Statist. Soc. Ser. C = Applied Statistics* **16**, 185–210 (with discussion).

Cohen, Morris R. and Nagel, Ernest (1934) *An Introduction to Logic and Scientific Method*. New York: Harcourt. (The unabridged edition.)

Cox, R. T. (1961) *The Algebra of Probable Inference*. Baltimore: John Hopkins University.

Deaton, M. L. and Foutz, R. V. (1980) 'Path analysis for complex-valued random variables with an application to spectral analysis of time series', C75 in *J. Statist. Comput. & Simul.* **12**, 61–65.

Edwards, A. W. F. (1963) 'The measure of association in a 2 × 2 table', *J. Roy. Statist. Soc. A*, **126**, 109–114.

Feynman, R. P. (1948) 'Space-time approach to non-relativistic quantum mechanics', *Reviews of Modern Physics* **20**, 367–387. Reprinted in *Quantum Electrodynamics* (J. Schwinger, ed.). New York: Dover, 1958, 321–341.

Feynman, R. P. and Hibbs, A. R. (1965) *Quantum Mechanics and Path Integrals*. New York: McGraw-Hill.

Fleiss, J. L. (1973) *Statistical Methods for Rates and Proportions*. New York: Wiley.

Good, I. J. (1959) 'Kinds of probability', *Science* **129**, 443–447. Reprinted in Good (1983b).

Good, I. J. (1961) 'A causal calculus', *Brit. J. Philos. Sci.* **11**, 305–318; **12**, 43–51; **13** (1962), 88. (Reprinted with the errata incorporated in Good, 1983b.)

Good, I. J. (general editor) (1962) *The Scientist Speculates: An Anthology of Partly-*

Baked Ideas. London: Heinemann; New York: Basic Books. Also German and French translations, Econ Verlag and Dunod respectively.

Good, I. J. (1965) 'Speculations concerning the first ultra-intelligent machine', *Advances in Computers* **6**, 31–88. (An index is available from the author.)

Good, I. J. (1966) 'A derivation of the probabilistic explication of information', *J. Roy. Statist. Soc. B*, **28**, 578–581.

Good, I. J. (1967) Contribution to the discussion of Buck and Wicken (1967).

Good, I. J. (1968) 'Corroboration, explanation, evolving probability, simplicity, and a sharpened razor', *Brit. J. Philos. Sci.* **19**, 123–143.

Good, I. J. (1969) 'Discussion of Bruno de Finetti's paper "Initial probabilities: A prerequisite for any valid induction",' *Synthese* **20**, 17–24. Reprinted in *Induction, Physics and Ethics: Proceedings and Discussion of the 1968 Salzburg Colloquium on the Philosophy of Science*, P. Weingartner and G. Zecha (eds.), Synthese Library, D. Reidel, Dordrecht, Holland, 1970, pp. 18–25.

Good, I. J. (1971) 'The probabilistic explication of information, evidence, surprise, causality, explanation, and utility', in *Foundations of Statistical Inference: Proc. Symp. on the Foundations of Statistical Inference*, V. P. Godambe and D. A. Sprott (eds.). Toronto: Holt, Rinehart, and Winston, pp. 108–141 (with discussion). Largely reprinted in Good (1983b).

Good, I. J. (1972a) 'Review' of Patrick Suppes, *A Probabilistic Theory of Causality*, *Acta Philosophica Fennica*, Fasc. **XXIV**. In *J. Amer. Statist. Assoc.* **67**, 245–246.

Good, I. J. (1972b) 'Random thoughts about randomness', *Philosophy of Science Assoc. 1972*. Boston Studies in the Philosophy of Science, Dordrecht: D. Reidel, 1974, pp. 117–135. Reprinted in Good (1983b).

Good, I. J. (1977) 'Explicativity: A mathematical theory of explanation with statistical applications', *Proc. Roy. Soc. (London)* **A354**, 303–330; **A377** (1981), 504. Partly reprinted in Good (1983b).

Good, I. J. (1978) 'Path analysis and correlation for power functions', C22 in *J. Statist. Comput. & Simul.* **8**, 80.

Good, I. J. (1979a) 'Fallacies, statistical', in *The International Encyclopedia of Statistics*, William H. Kruskal and Judith M. Tanur (eds.). New York: The Free Press, pp. 337–349.

Good, I. J. (1979b) 'Turing's statistical work in World War II' (Studies in the history of probability and statistics, vol. XXXVII.), *Biometrika* **66**, pp. 393–396.

Good, I. J. (1980a) 'Some comments on probabilistic causality', *Pacific Philosophical Quarterly* **61**, 301–304.

Good, I. J. (1980b) 'A further comment on probabilistic causality: Mending the chain', *Pacific Philosophical Quarterly* **61**, 452–454.

Good, I. J. (1980c) 'Degrees of causation in regression analysis', C71 in *J. Statist. Comput. & Simul.* **11**, 153–155.

Good, I. J. (1980d) 'Feynman's path integrals and Sewall Wright's path analysis', C80 in *J. Statist. Comput. & Simul.* **12**, 74–77.

Good, I. J. (1980e) 'A simplification in the "causal calculus"', C81 in *J. Statist. Comput. & Simul.* **12**, 77–78. (Reprinted in Good, 1983b.)

Good, I. J. (1981) 'The effect of permutations of rows and columns on measures of association', C103 in *J. Statist. Comput. & Simul.* **13**, 309–312. (In the last three lines of page 309, *b* and *c* have been interchanged in error.)

Good, I. J. (1983a) Review of *The Maximum Entropy Formalism*, Raphael D. Levine and Myron Tribus (eds.) (1979), *J. Amer. Statist. Assoc.* **78**, 987–989.

Good, I. J. (1983b) *Good Thinking: The Foundations of Probability and Its Applications*. University of Minnesota Press.

Good, I. J. (1984) 'The best explicatum for weight of evidence', C197 in *J. Statist. Comput. & Simul.* **19**, 294–299.

Good, I. J. (1985a) 'Weight of evidence: A brief survey', in *Bayesian Statistics 2: Proceedings of the Second Valencia International Meeting, September 6/10, 1983*), J. M. Bernardo, M. H. DeGroot, D. V. Lindley, and A. F. M. Smith (eds.). New York: North Holland, pp. 249–269 (including discussion).

Good, I. J. (1985b) 'Causal propensity: A review', in *PSA 1984*. Volume 2.

Good, I. J. (1985c) 'Explanatory power depends on more than probabilities', C238 in *J. Statist. Comput. & Simul.* **22**, No. 2, 184–186.

Good, I. J. (1985d) 'Some statistical applications of Poisson's work', *Statistical Science*. **1** (1986), 157–180.

Good, I. J. (1985e) 'A correction concerning degrees of causation', C218 in *J. Statist. Comput. & Simul.* **20**, 326.

Good, I. J. (1985f) 'A reinstatement, in response to Gilles, of Redhead's argument in support of induction', *Philosophy of Science* **54** (1987), 470–472.

Good, I. J. and McMichael, Alan F. (1984) 'A pragmatic modification of explicativity for the acceptance of hypotheses', *Philosophy of Science* **51**, 120–127.

Good, I. J. and Mittal, Y. (1985) 'The amalgamation and geometry of two-by-two contingency tables', *Annals of Statistics* **15** (1987), 694–711.

Goodman, L. A. and Kruskal, W. H. (1954, 1959) 'Measures of association for cross-classifications', *J. Amer. Statist. Assoc.* **49**, 732–764; **54**, 123–163. Reprinted in a book of the same title. New York: Springer, 1979.

Granger, C. W. J. and Hatanaka, M. (1964) *Spectral Analysis of Economic Time Series*. Princeton, N.J.: Princeton University Press.

Greenberg, B. G. (1969) 'Problems of statistical inference in health with special reference to the cigarette smoking and lung cancer controversy', *J. Amer. Statist. Assoc.* **64**, 739–758.

Hart, H. L. A. and Honoré, A. M. (1959) *Causation in the Law*. Oxford: Clarendon Press.

Humphreys, Paul (1980) 'Cutting the causal chain', *Pacific Philosophical Quarterly* **61**, 305–314.

Keynes, J. M. (1921) *A Treatise on Probability*. London: Macmillan.

Lindley, D. V. and Novick, M. R. (1981) 'The role of exchangeability in inference', *Annals of Statistics* **9**, 45–58.

Ogden, C. K. (1932) *Bentham's Theory of Fictions*. London: Routledge & Kegan Paul.

Reichenbach, Hans (1956) *The Direction of Time*. Berkeley: University of California Press.

Royal Corps of Signals (1947) *Handbook of Line Communication*, Volume 1. London: His Majesty's Stationery Office.

Russell, Bertrand (1948) *Human Knowledge*. London: George Allen & Unwin.

Salmon, W. (1980) 'Probabilistic causality', *Pacific Philosophical Quarterly* **61**, 50–74. (Raises some objections to Good, 1961, which are answered in Good, 1980a, but not all correctly.)

Savage, L. J. (1954) *Foundations of Statistics*. New York: Wiley.

Sheps, M. C. (1958) 'Shall we count the living or the dead?' *The New England J. of Medicine* **259**, 1210–1214.

Sheps, M. C. (1959) 'An examination of some methods of comparing several rates or proportions', *Biometrics* **15**, 87–97.

Simpson, E. H. (1951) 'The interpretation of interaction in contingency tables', *J. Roy. Statist. Soc. B*, **13**, 238–241.

Suppes, Patrick (1970) *A Probabilistic Theory of Causality*. Amsterdam: North-Holland (Acta Philosophica Fennica, Fasc. XXIV).

Tate, Robert F. (1978) 'Path analysis', in *International Encyclopedia of Statistics*, Volume 1, W. H. Kruskal and J. M. Tanur (eds.). New York: The Free Press, pp. 624–628. (Part of his article on correlation methods.)

Vaihinger, Hans (1911) *Die Philosophie des Als Ob. System der theoretischen, praktischen und religiosen Fiktionen der Menscheit auf Grund eines idealistischen Positivismus. Mit ein Anhang über Kant und Nietzsche*, 6th edn. F. Meiner, Leipzig (1920). Parts of this sixth edition were translated by C. K. Ogden in *The Philosophy of 'As If': A System of Theoretical, Practical and Religious Fictions of Mankind*, 2nd edn. London: Routledge & Kegan Paul (1935).

Wiener, Norbert (1956) 'The theory of prediction', in *Modern Mathematics for the Engineer*, E. F. Beckenbach (ed.). New York: McGraw-Hill, pp. 165–190.

Wright, Sewall (1918) 'On the nature of size factors', *Genetics* **3**, 367–374.

Yule, G. U. (1903) 'Notes on the theory of association of attributes in statistics', *Biometrika* **2**, 121–134. Reprinted in *Statistical Papers of George Udney Yule* (London: Griffin), 71–84.

Yule, G. U. (1911) *An Introduction to the Theory of Statistics* (London: Griffin). (The 11th edn. of 1937 and all later editions, were authored by Yule and M. G. Kendall.)

WESLEY C. SALMON

INTUITIONS – GOOD AND NOT-SO-GOOD

In (Salmon, 1980) I offered a critical survey of what I regarded as the three significant theories of probabilistic causality available at that time – namely, those of I. J. Good, Hans Reichenbach, and Patrick Suppes. Both Good and Suppes have responded to that article – (Good 1980 and 1985), and (Suppes 1984, chap. 3) – and I have taken the subject up again in (Salmon 1984, chap. 7). The purpose of this paper is to continue that discussion. As I see it, we have arrived at a point at which basic intuitions about probabilistic causality clash.[1] It may not be possible to resolve the conflicts, but I hope at least to clarify the issues.

Good, in his (1961–62), set forth an ambitious and complicated quantitative theory of probabilistic causality – the kind of theory which, if sound, would certainly be wonderful to have. He defined two measures – $Q(E:F)$, the degree to which F tends to cause E; and $\chi(E:F)$, the degree to which F actually causes E. My first criticism was that Q is nothing but a measure of statistical relevance, and that it fails to distinguish between what Suppes (1970) called genuine causes and spurious causes. That criticism was based on a misunderstanding, as (Good 1980) pointed out; for this I want simply to retract the criticism and apologize to him. Moreover, even if I had been right in my interpretation, it would have been a simple matter to repair the defect.

Another issue raised in (Salmon 1980) is not as easily resolved. In Good's causal calculus, the strength S of a causal chain is a function of the Q-values of its links. I offered two counterexamples designed to show that this functional relation cannot hold. In his (1980), Good responded that he simply did not see in what way my examples support my case. I had thought that the force of the counterexamples was intuitively obvious. But then, as we all know, one person's counterexample is another person's *modus ponens*. So let's look at the counterexamples.

The first putative counterexample involves a comparison between two simple games. In the first game, the player begins by tossing a tetrahedron with sides marked 1, 2, 3, 4 (event F). If the tetrahedron

Brian Skyrms and William L. Harper (eds.), Causation, Chance, and Credence, Vol. I, 51–71.

comes to rest with side 4 showing, the player draws from a deck containing 16 cards, 12 red and 4 black (event G). If the tetrahedron comes to rest with side 4 on the bottom, the player draws from another deck containing 16 cards, 4 red and 12 black (event $\bar{G}$). Drawing a red card constitutes a win (event E). In this game the probability of winning on any play is 10/16 (see Figure 1).

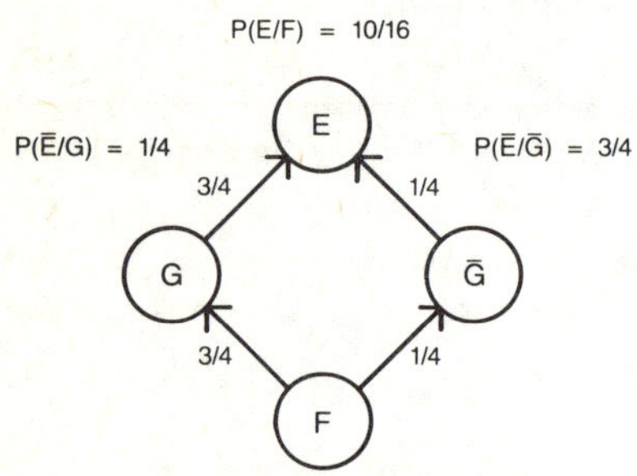

Tetrahedron - card game
First game

Psychotherapy
Joe Doakes

Figure 1a

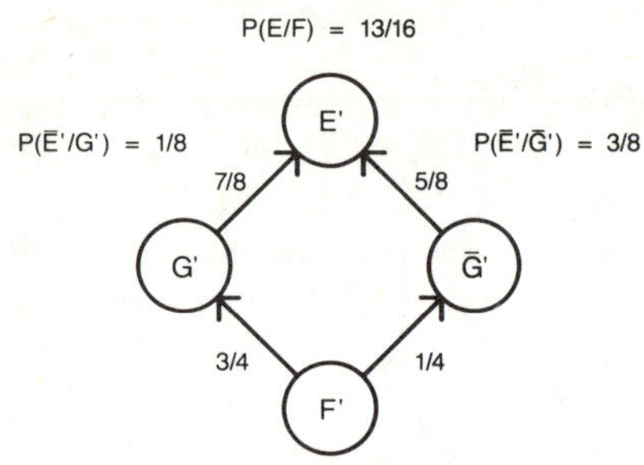

Tetrahedron - card game
Second game

Psychotherapy
Jane Bloggs

Figure 1b

The second game is just like the first except for the makeup of the two decks. In this game the first deck contains 14 red and 2 black cards; the other deck contains 10 red cards and 6 black. The events in this game are designated by primed letters. In this game the probability of winning on any play is 13/16 (see Figure 1).

Suppose, now, that one play of each game has occurred, and in each case the 4 showed on the toss of the tetrahedron, with the result that the player drew from the favorable deck. Moreover, each player drew a red card. Thus, we have two chains of events: $F \rightarrow G \rightarrow E$ and $F' \rightarrow G' \rightarrow E'$.

In that paper I argued as follows. According to Good's definition,

$$\begin{aligned} Q(Y{:}X) &= \log[P(\bar{Y}|\bar{X})/P(\bar{Y}|X)] \\ &= \log\{[1 - P(Y|\bar{X})]/[1 - P(Y|X)]\}. \end{aligned}$$

On the reasonable supposition that the only way a player gets to draw from either deck in either game is by entering and tossing the tetrahedron, we have

$$P(G|F) = P(G'|F') = 3/4; \qquad P(G|\bar{F}) = P(G'|\bar{F}') = 0$$

Since

$$[1-P(G|\bar{F})]/[1-P(G|F) = [1-P(G'|\bar{F}')]/[1-P(G'|F')],$$

it follows that

$$Q(G:F) = Q(G':F') \qquad (= \log 4)$$

From the given probability values we calculate that

$$\begin{aligned} Q(E:G) &= \log\{[1 - P(E|\bar{G})]/[1 - P(E|G)]\} \\ &= \log[(3/4)/(1/4)] = \log 3 \end{aligned}$$

and

$$\begin{aligned} Q(E':G') &= \log\{[1 - P(E'|\bar{G}')]/[1 - P(E'|G')]\} \\ &= \log[(3/8)/(1/8)] = \log 3 \end{aligned}$$

Therefore,

$$Q(E:G) = Q(E':G').$$

Consequently, on Good's measure S of strength – according to which the strength of a causal chain is a function of the Q-values of its links – the causal strength of the chain $F \to G \to E$ equals that of $F' \to G' \to E'$. Comparing the two games, it seemed intuitively evident to me that the strength of the primed chain is greater than that of the unprimed chain. After all, E can come about in either of two ways and E' can come about in either of two ways, and for each way the probability of E' is greater than the probability of E. In addition, given our stipulation that the only way a win can occur in either game is by tossing the tetrahedron and drawing from the appropriate deck, the statistical relevance of F' to E' is greater than that of F to E. Good informed me in a letter (15 June 1979) that he was unmoved by this example.

In an attempt to make my objection more compelling, I offered a second putative counterexample involving two cases. Although there are fundamental differences between the first and second examples, the causal chains have the same associated probabilities in both examples (see Figure 1).

> Suppose that two individuals, Joe Doakes and Jane Bloggs, suffer from sexual dysfunctions. Joe is impotent and Jane is frigid. Each of them decides to seek psychotherapy. There are two alternative types of therapy available, directive and nondirective. When Joe seeks out a psychotherapist (event F), there is a probability of 3/4 that he will select a directive therapist and undergo that type of treatment (event G), and a probability of 1/4 that he will select a nondirective therapist and undergo that type of treatment (event $\bar{G}$). If he is treated by a directive therapist, there is a probability of 3/4 that he will be cured (event E), and if he is treated by a nondirective therapist, there is a probability of 1/4 that he will be cured. Given these values, there is a probability of 10/16 that he will be cured, given that he undertakes psychotherapy.
>
> When Jane seeks out a psychotherapist (event F'), there is a probability of 3/4 that she will select a directive therapist (event G'), and a probability of 1/4 that she will select a nondirective therapist (event $\bar{G}'$). If she is treated by a directive therapist, there is a probability of 7/8 that she will be cured (event E'), and if she is treated by a nondirective therapist, the probability of a cure is 5/8. Given these values, there is a probability of 13/16 that she will be cured, given that she undertakes psychotherapy.
>
> Joe and Jane each undertake psychotherapy, each is treated by a directive therapist, and each is cured. [They meet, fall in love, and live happily ever after.] Thus, we have two causal chains, $F \rightarrow G \rightarrow E$ and $F' \rightarrow G' \rightarrow E'$. . . . The question is, on what basis, if any, would we be warranted in claiming that the two chains have the same strength – i.e., that the degree to which the seeking out of psychotherapeutic treatment caused the cure is the same for both? (Salmon 1980, p. 54.)

In this example, the initial cause F is the decision to undertake psychotherapy. It doesn't involve the selection of a therapist and the choice of a particular type of treatment. Ann Landers frequently advises people to "get counselling," without specifying anything about the type. Indeed, the advice, often, is to talk to a clergyman, a family doctor, a psychologist, a psychiatrist, or a social worker. It's highly unspecific. The idea behind my example is to ask how efficacious in overcoming either of the dysfunctions is the taking of such general advice as "get psychotherapy." In this example I intend to confine the types of treatment considered to those offered by psychiatrists or clinical psychologists. Arbitrarily, perhaps, results of anything else fall into the category of spontaneous remission.

According to Good's causal calculus the strengths of the two chains must be equal, for the same reasons as in the first example. He apparently finds this conclusion unobjectionable. I suggested that we cannot answer the question about the causal efficacy of psychotherapy in these two cases until we know the probabilities of spontaneous remission for the two conditions. Let us therefore arbitrarily stipulate some values: 1/100 for impotence and 3/4 for frigidity. Given these values, I claimed that the strength of the causal chain in Joe's case is greater than it is in Jane's, since the probability of a cure with the aid of psychotherapy for Joe (10/16) is much greater than the rate of spontaneous remission, while the probability of a cure with the aid of psychotherapy for Jane (13/16) is only slightly greater than the spontaneous remission rate. We should not be misled by the words "cure" and "remission"; each refers simply to the fact that the problem went away. There is no reliable way to say in advance whether the cure will last. We cannot legitimately construe "cure" to mean something permanent and "remission" to refer to a temporary condition that will be followed by a recurrence of the problem.

Good was no more moved by the second example than by the first. He remarks, "I do not see in what way his examples support his case except that what is said three times sounds true" (1980, p. 302). He suggests that my belief in the difference in strength of the two chains results from confusing Q (the tendency to cause) with χ (the strength of the causal connection). The Q-values are different: $Q(E:F) = 0.971$ while $Q(E':F') = 0.288$.[2] We both agree that the tendency of psychotherapy to eliminate a sexual dysfunction is different for Bloggs and Doakes, but we disagree about the degree to which psychotherapy actually caused the cure in the two cases. I want to try to understand this disagreement.

In an effort better to understand Good's theory – in particular, the distinction between Q and χ – let us now consider some of the examples he has offered. In (1961–62) Good gives an example to show that Q and χ cannot be identified:

> Sherlock Holmes is at the foot of a cliff. At the top of the cliff, directly overhead, are Dr Watson, Professor Moriarty, and a loose boulder. Watson, knowing Moriarty's intentions, realises that the best chance of saving Holmes's life is to push the boulder over the edge of the cliff, doing his best to give it enough horizontal momentum to miss Holmes. If he does not push the boulder, Moriarty will do so in such a way that it will be nearly certain to kill Holmes. Watson then makes the decision (event F) to push the boulder, but his skill fails him and the boulder falls on Holmes and kills him (event E).
>
> This example shows that $Q(E:F)$ and $\chi(E:F)$ cannot be identified, since F had a tendency to prevent E and yet caused it. We say that F was a cause of E because there was a chain of events connecting F to E, each of which was strongly caused by the preceding one. (Good 1961–62, pt. I, p. 318.)

This example, which I have reproduced in its entirety, is the only concrete illustration furnished in (Good 1961–62) of the difference between Q and χ. While it has some heuristic value, it suffers from the fact that degrees of causal efficacy are not involved. Watson's decision fully failed to prevent E; indeed, it fully caused E, given the conditions provided in the example. So it doesn't help much in trying to understand the Bloggs–Doakes example. The fact that Moriarty would have killed Holmes if Watson had done nothing has no bearing on the fact that Watson killed Holmes. One important difference between this example and the Bloggs–Doakes example is the fact that the cause of Holmes's death is readily identifiable as an act of Watson rather than an act of Moriarty. The Bloggs–Doakes example defies such an analysis.

In (Good 1980) another example is offered to illustrate the distinction between Q and χ, namely, the distinction between murder and attempted murder. It is easy to see how an attempt at murder has a tendency to cause the death of the victim, and to see how the strength of that tendency might be quantified in degrees, depending on the skill and motivation of the perpetrator, the conditions under which the attempt is made, and steps taken to prevent the murder. If, however, the attempt is successful and the victim is killed, it is not easy to see how to assign a degree to the contribution of the murderer to the death of the victim other than to say simply that the murderer did it. Even if the murderer is a professional assassin, and even if the person who took out the contract on the life of the victim hires several other professional assassins as well to make sure that the victim dies, it is hard to see how – if they all work independently – the successful assassin should fail to get full credit and the others no credit whatever.

One complication that might arise, of course, is for two assassins to shoot the victim at precisely the same moment, each shot being sufficient to cause immediate death by itself. This makes the example the same in principle as another of Good's examples, namely, the firing squad. We shall look at that example in a moment. Such examples are cases of overdetermination, about which much has been written. As we shall see, Good has an interesting method for dealing with them.

The situation would be somewhat more complicated if two assassins were to work together. Suppose, for example, that the intended victim is in a building with only two exits. One assassin enters through the front door; the other waits outside of the back door. As the killers arrive on the scene, it is impossible to predict which one will actually kill this victim. If he remains in his office he will confront the first assassin and be killed by him; if he makes a run for it through the back door the other assassin will kill him. In this example, both commonsense and the law hold both assassins responsible. I'm not at all sure how we should quantitatively apportion the blame. If, in this case, both were indispensable, then perhaps it is reasonable to divide the responsibility equally between them. If, however, there is some small chance that the assassin who enters the front door would be successful without the aid of the other assassin, while the one waiting outside of the back door has no chance of success working by himself, then it would probably be reasonable to assign to the first assassin a higher causal contribution than to the second.

Where basic intuitions seem to conflict, it is useful – indeed, indispensable – to consider a variety of examples. Prior to Good's PSA paper (Good, 1985), he had furnished a paucity of examples to illustrate his basic concepts; in fact, the only ones with which I was familiar were the little Sherlock Holmes story and the unelaborated mention of the distinction between murder and attempted murder. Fortunately, his PSA paper as well as his principal contribution to this volume contain a number of useful ones.

One of his examples, which follows naturally from our elaboration of the distinction between attempted murder and murder, involves a firing squad with two marksmen, both crack shots. When the captain gives the order to fire (event F), both shoot (events G_1 and G_2). Each shot would, by itself, be sufficient to cause death (event E). In this case there are two causal chains from F to E, and each has maximal strength (positive infinity). Again, we get no real feeling from such examples as to the way to assign nonextreme degrees of causation.

That drawback is overcome in an ingenious variation on the firing squad example. Suppose the squad has three members, all crack shots, each of whom uses a standard six-shot revolver. When the captain gives the order to fire, each marksman spins the cylinder of his weapon (as in Russian roulette), aims at the condemned, and pulls the trigger. If a bullet is in the firing position when the trigger if pulled, a fatal shot is fired. The condemned person dies if and ony if at least one shot is fired – where it is possible, of course, that more than one shot is fired. In this case we have more than one causal chain, and each contains links of nonextreme strength. It provides a good example of a causal net.[3]

Now that we have before us several examples (due to Good or

myself) that have appeared in the literature, let us attempt to examine them in a more systematic manner. At the outset a simple but basic distinction must be made. There are cases in which two or more causes acting together bring about a result – e.g., reaching the goal in a fund drive is achieved because gifts are received from many sources, and each gift constitutes a definite portion of the amount raised. Here it is natural to quantify the degree to which each cause contributed to the effect. There is, of course, nothing probabilistic about this example.

In a different sort of case we have two or more causes, any one of which might, by itself, bring about a result. The potential causes are mutually exclusive; if one operates the others do not. For instance, on any given morning I may walk to my office, drive, or take a bus. When I adopt one mode of transport I reject the others for that day. This example is also nonprobabilistic.

The Russian roulette firing squad exemplifies the first situation probabilistically. There are three intermediate causes that may be said to cooperate in bringing about a result; all three marksmen are present and follow orders for any given "execution" (see Figure 2). As I construe this example, the condemned person may survive the carrying out of the sentence, because the order to fire is given only once for any person sentenced to face that sort of firing squad. Thus, neither singly nor in concert do they inevitably produce death. As we shall see shortly, this example lends itself nicely to quantitative analysis.

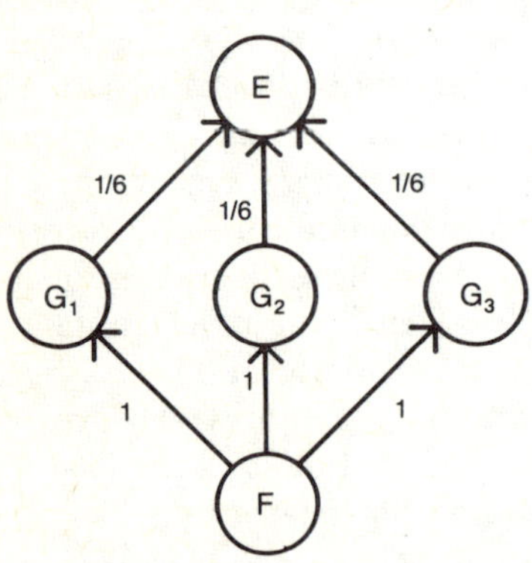

Russian roulette firing squad

Figure 2a

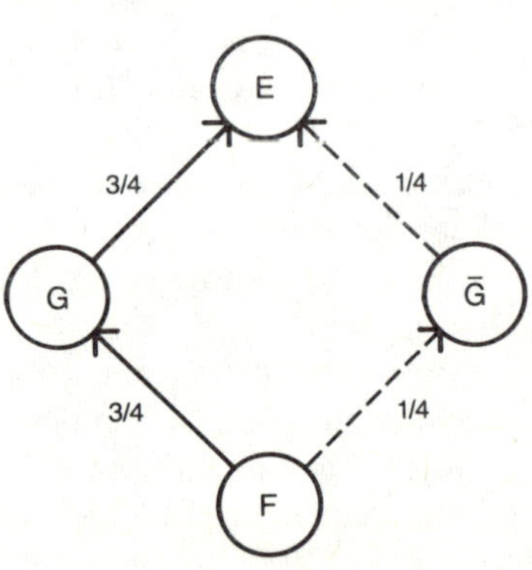

Psychotherapy

Joe Doakes

Figure 2b

Some of our examples fit the pattern of several different alternative causes, only one of which is present in any given case (see Figure 2). For instance, both versions of the tetra-card game exemplify it, for on each play of both games the player draws from one deck only. Good's example of the pinball machine also fits this case, for there are many possible routes a ball can travel before it drops out of play at the bottom. On any given play, of course, it follows only one route. The two versions of the psychotherapy example may also seem to fit, for Bloggs and Doakes each undergo only one form of therapy. It is probably more realistic, however, to consider the factors – whatever they may be – that sometimes lead to spontaneous remission to be additional probabilistic causes that are also present when a person undergoes psychotherapy. It would be implausible to suppose that the decision to undertake psychotherapy eliminates them.

Let us see how situations of both types are handled. Good has remarked on several occasions that an analogy between causal nets and electrical circuits has provided fruitful intuitions. Indeed, he introduces the notions of resistance R and strength S of the elements of causal nets (where strength is analogous to electrical conductance). In circuit theory resistance and conductance are reciprocal; in Good's causal calculus their counterparts are not quite reciprocal, but as one increases the other decreases in a precisely defined way.[4]

Let us look at the analogy. In circuit theory the resistance of a circuit composed of several elements connected in series is equal to the sum of their resistances. The conductance of a circuit composed of several elements connected in parallel is equal to the sum of their conductances. These relations are intuitively obvious – as well as being experimentally confirmed (see Figure 3). If a circuit has three parallel paths of equal conductance C, the conductance of the circuit is $3C$. Exploiting the well-worn analogy between electricity and hydraulics, we may say that three times as much "juice" can flow through all three channels as can flow through one alone. Indeed, the electrical current consists of a flow of electrons. Thinking now of a causal net, the strength of the causal connection in a net containing three chains of strength S in parallel should be equal to $3S$, the sum of the strengths of three chains. Good stipulates, however, that the chains must be mutually independent (1961–62, pt. I, pp. 313–14).

Let us apply these ideas to some examples. The simple firing squad with two members who fire fatal shots if and only if the captain gives

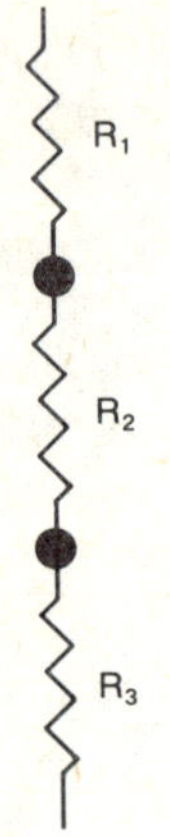

$R = R_1 + R_2 + R_3$

Resistors in series

Figure 3a

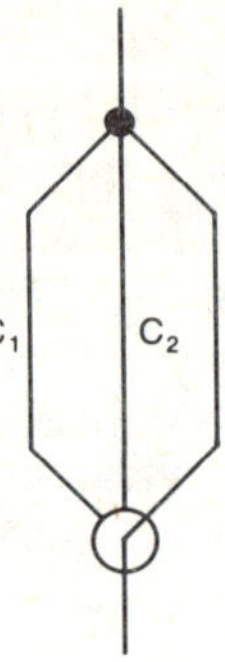

$C = C_1 = C_2 = C_3$

Conductors in parallel
(Connected only one at a time)

Figure 3b

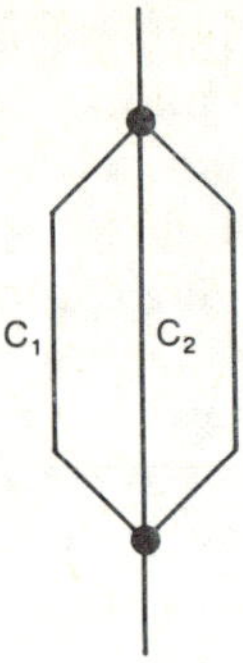

$C = C_1 + C_2 + C_3$

Conductors in parallel

Figure 3c

the command to fire is a net consisting of two chains in parallel. Since each probability is unity, all of the links have positive infinity as their degrees of strength. Since the strength of the net is equal to the sum of the strengths of the chains, the strength of the net is also positive infinity. Overdetermination thus poses no problem for the causal calculus.

Let us now consider a nondegenerate example. In the case of the Russian roulette firing squad, event F, the command to fire, gives rise to G_1, G_2, and G_3 with a probability of one. Consequently, the links between F and each of the three have maximal strength of positive infinity. Each event G_i has a probability of 1/6 to produce a fatal shot. At this point we must exercise a bit of care. If we take E to stand for the death of the condemned person, and if we compute $P(E|G_i)$, we must recall that, if one marksman pulls the trigger, all three do. Thus, the probability that the condemned dies if one marksman pulls the trigger is the probability that at least one of the three fires a fatal shot, i.e.,

$$P(E|G_i) = 1 - (5/6)^3 = 0.421$$

Clearly, this is the wrong probability to use to evaluate the causal contribution of a given marksman to the death of the condemned, for the fact that one marksman pulls the trigger makes no causal contribution to the fact that the others do also – it is the command of the commissar that causes each to pull his or her trigger. It would be an obvious mistake to use that value to calculate the strength of one of the chains, and then to multiply by three to get the strength of the entire net. Instead, we might write E_i to signify that a fatal shot was fired by G_i. E is then equivalent to the nonexclusive disjunction of the E_i. We calculate the Q-value of each link as follows:

$$Q(E_i|G_i) = \log\{[1 - P(E_i|\bar{G}_i)]/[1 - P(E_i|G_i)]\} = 0.182$$

In Good's causal calculus the strength of a single link is equal to the Q-value of that link, so $S(E:G_i) = 0.182$. Given the maximal value of the first link, the strength of any one of the three chains $F \rightarrow G_i \rightarrow E_i$ is also 0.182. The strength of the entire net is the sum of the strengths of the three chains connected in parallel, namely, 0.547. That value represents $\chi(E:F)$, the actual causal contribution of the commissar's order "Fire!" to the death of the condemned person. In this case, it happens, $Q(E:F)$, the tendency of the captain's order to cause death, has the same value, as can be shown by computing the value of log $\{[1 - P(E|\bar{F})]/[1 - P(E|F)]\}$.

Let us also look at the case of Jane Bloggs. Using the probability values as given, we can easily compute $Q(G':F') = \log 4 = 1.39$ and $Q(E':G') = \log 3 = 1.10$.[5] For the separate links, $S(G':F') = Q(G':F')$ and $S(E':G') = Q(E':G')$. Since the two links are connected in series, we can get the resistance of the chain by computing the sum of the resistances of the links. Strength and resistance are related by the formulas

$$R(Y:X) = -\log[1 - e^{-S(Y:X)}]$$
$$S(Y:X) = -\log[1 - e^{-R(Y:X)}]$$

According to my calculations,

$$R(G':F') = 0.405; \qquad R(E':G') = 0.288;$$

which yields

$$S(E':F') = 0.693$$

as the strength of the chain $F' \rightarrow G' \rightarrow E'$.

Let us now consider Jane's cousin Amy who suffered from the same sexual dysfunction as Jane, undertook psychotherapy, and got over her frigidity. In contrast to Jane, however, Amy sought help through nondirective therapy. In her case we have the chain $F' \rightarrow \bar{G}' \rightarrow E'$; we find that $Q(\bar{G}':F') = 0.288$, but $Q(E':\bar{G}')$ has a negative value (-1.10). Any link that has a negative Q-value has, by definition, a strength of zero, and any chain that has such a link also has zero strength. In Good's calculus, no chain is stronger than its weakest link. If we consider only the case of Jane Bloggs, this result may not seem troubling. But what are we to say of Amy? Is there, in her case, no causal chain? The intuitive answer is that there is indeed a causal chain. We shall have to develop a sounder method for evaluating cases of this sort.[6]

The question just raised is the problem of negative relevance, which was vividly posed in (Suppes 1970) in terms of Deborah Rosen's famous case of the spectacular birdie.[7] It is further discussed in (Salmon 1980, 1984) and (Suppes 1984). A similar example is given by Good (1985). In both examples a skillful golfer makes a shot, with the intention of getting the ball in the hole (Rosen) or on the green (Good). In Rosen's example, the shot is actually quite poor, but the ball hits a tree branch and is deflected into the hole. In Good's example, the tee shot might have been good or poor (he doesn't say which), but (in either case) the ball strikes a bird, falls to the ground, where it is picked up by a chimpanzee who carries it to the green and drops it there. That's called "making it the hard way."[8]

Good uses this example to illustrate the distinction between Q and χ; he remarks, "Here $Q(E:F)$ is appreciable but $\chi(E:F)$ is negligible." Why is the Q-value appreciable? Perhaps because this particular shot is excellent; but for the bird it was almost certain to land on the green. Why is the χ-value negligible? Because the causal chain from stroke to bird to chimp to green has some very weak links. When the ball hits the bird the probability that it will get to the green is tiny, for it is most unlikely that there will be a chimp or other messenger there to get it from the spot at which it landed to the green. Moreover, when the chimp picks up the ball the probabilities may overwhelmingly favor the chimp carrying the ball back to the language lab rather than depositing it on the green. In Good's causal calculus a theorem states that no causal chain is stronger than its weakest link, and this chain has some very weak links.

This example seems to illustrate quite well (qualitatively) the distinction between Q and χ, but there is one major question: is it a causal chain? To qualify as a causal chain a series of events must sustain suitable relations of positive relevance to one another; in particular, each event in the series must be positively relevant to its immediate successor. The first event in Good's example is the tee stroke, and the second is the striking of the bird. It is reasonable to say that the ball is more likely to hit the bird if it is driven from the tee than if it is not. The third event is the landing of the ball (somewhere on the fairway or in the rough). I find it difficult to believe that the probability of the ball landing in about the same place, given that it hits the bird, is not pretty small. That event must be quite sensitive to the position and attitude of the bird, and, for a bird in flight, those parameters change very quickly and quite irregularly. But the question is not whether that probability value is small; it is a question of relevance. So we must ask, what if the ball had not struck the bird? Well, if the immediately preceding event is the tee stroke, then one might suspect that it is more probable that the shot from the tee would put it there than that it would fall in that general location given that it has struck the bird. Of course we lack statistics on results of collisions of golf balls with flying birds (and experiments would obviously be difficult to conduct), so I could be wrong about this matter.

Given that the ball drops to the ground where it does, it will take a miracle to get it to the green before it comes to rest, but we are not disappointed: enter the chimp. The next event is either the entry of the chimp or the chimp's picking up the ball. Either way we seem to be in trouble, for the striking of the bird by the ball is certainly irrelevant to the presence of the chimp in that general vicinity, and it is also irrelevant to the chimp's picking up the ball. If the chimp has picked up the ball, is that positively relevant to the later presence of the ball on the green? The answer must be affirmative, for given the place the ball has reached after colliding with the bird, there isn't any other way the ball will get to the green. So even if the probability of the chimp carrying the ball to the green is not large – it's much more likely that he will carry it back to the language lab – it's still positively relevant.

In my (1980, pp. 62–65), in discussing several examples of the foregoing kind that seem to involve chains with negatively relevant links, I formulated two (not mutually exclusive) strategies that have been adopted by various authors. I called them *the method of inter-*

polated causal links and *the method of more detailed specification of events.* I tried to show that they are inadequate to restore positive relevance to all links of causal chains. I suggested (Salmon 1984, pp. 196–202) that an approach called *the method of successive reconditionalization* would be more successful. Although Good's approach, which conditionalizes not only on the given event but also on the state of the universe just prior to that event, does not embody precisely this method of circumventing the problem of negative relevance, it can easily be modified so as to do so.

In his earliest presentation of the causal calculus, Good emphasized that the notation we have been using thus far is condensed, and that all probabilities are to be taken conditional upon the state of the universe U and all of the laws of nature H. Hence, $P(G|F)$ should be construed as $P(G|F \,\&\, H \,\&\, U)$, where U is the state of the universe just before F occurs; similarly, $P(E|G)$ stands for $P(E|G \,\&\, H \,\&\, U')$, where U' is the state of the universe just before G occurs. It was because of my failure fully to appreciate the import of this interpretation of the probability expression that I failed to realize in (1980) that Good does have the distinction between genuine and spurious causes built in. A small (but important) modification would make it possible for his causal calculus to incorporate the method of successive reconditionalization as well. What is required is to stipulate that the probabilities are to be conditioned on the state of the universe at the time of the event in question instead of just prior to its occurrence.

For purposes of illustration let us apply this approach to the Bloggs case. Our previous calculations of $Q(G':F')$ and $Q(\bar{G}':F')$ can stand, for they were conditioned on the fact that F' actually obtained. When, however, we turn to $Q(E':G')$ and $Q(E':\bar{G}')$, we find that we must take care. In particular, when we calculate

$$Q(E':G') = \log\{[1 - P(E'|\bar{G}')]/[1 - P(E'|G')]\}$$

we must stipulate that $P(E'|\bar{G}') = 0$, for, given that G' obtains and $\bar{G}'$ does not, no E' can result from $\bar{G}'$. Similarly, when we calculate

$$Q(E':\bar{G}') = \log\{[1 - P(E'|G')]/[1 - P(E'|\bar{G}')]\}$$

we must set $P(E'|G') = 0$ for the analogous reason.[9] With this understanding we can properly calculate the strength of the causal chains of the Bloggs cousins. For Jane we have

$$Q(G':F') = 1.39; \qquad R(G':F') = 0.286$$
$$Q(E':G') = 2.08; \qquad R(E':G') = 0.133$$

which yields

$$S(F' \rightarrow G' \rightarrow E') = 1.07$$

For Amy we have

$$Q(\bar{G}':F') = 0.288; \qquad R(\bar{G}':F') = 1.39$$
$$Q(E':\bar{G}') = 0.981; \qquad R(E':\bar{G}') = 0.470$$

which yields

$$S(F' \rightarrow \bar{G}' \rightarrow E') = 0.169$$

Notice that positive relevance now obtains within this chain. Clearly, conditioning on the state of the universe at the time $\bar{G}'$ occurs does not affect the value of $P(E' \mid \bar{G}')$, but it does change the value of $P(E' \mid G')$, which occurs explicitly in the calculation of $Q(E':\bar{G}')$. There might be some temptation to add together the strengths of the two chains in order to calculate the strength of the entire net containing F', G', $\bar{G}'$, and E', but this temptation should be resisted, inasmuch as the two chains, being mutually exclusive, are not independent. It would be highly counterintuitive, to my mind, to assign the same strengths to the causal nets for Jane and Amy.

The method just outlined goes far in handling the problem of negative relevance for a number of problematic cases. However, as I pointed out in (Salmon 1980, pp. 65, 69–70) and (Salmon 1984, pp. 200–202), a putative counterexample from atomic physics apparently cannot be handled successfully by that method. Good (1980, p. 303) has denied the pertinence of this example on the ground that it is noncausal. I disagree with that claim (Salmon 1984, pp. 201–202), but I shall not say more about it here.

If we want to pursue the analogy with circuit theory, it seems crucial to distinguish two types of causal nets. Good defines a causal net as a collection of events, some actual and others perhaps only possible, in which is embedded at least one causal chain of actual events connecting the initial event F to the final event E. In some cases the net may contain a single actual chain – e.g., the Bloggs and Doakes cases, my tetra-card games, Good's pinball example – where there are other possible chains that are actualized (one at a time) in other similar situations, but they are all mutually exclusive. In other cases – e.g., Good's firing squad examples – there are several independent chains composed of actual or possible events connecting the initial event (the order "Fire!") to the final event (the death of the condemned person). The Russian roulette firing squad example, in which more than one chain of actual events linking the initial and final stages is possible, is analogous to the circuit in which all three channels are open simultaneously. Good's skin cancer example illustrates the same sort of situation. Examples of the first type, with mutually exclusive possible chains, are analogous to circuits containing two or more conductors connected by switches that allow only one channel to be connected at a time (see Figure 3). The conductance in this case is not the sum of the several conductances; it is simply the conductance of the conductor that happens to be connected at any given time. Where the possible chains are mutually exclusive, the merely potential conductors have no effect whatever on the conductance of the circuit.

The electrical circuit analogy is heuristically useful, I believe, in the case of nets containing mutually exclusive chains and for those containing mutually independent chains. Other cases are, of course, possible. One chain may be positively or negatively relevant to another; in such cases it would presumably be necessary to add strength or resistance, respectively, to the alternative possible chains. As we shall see below, the psychotherapy example may be a case in point. I do not have a concrete proposal for dealing with them, and it is not clear to me whether the circuit analogy is helpful these instances.

We must now turn to one final fundamental distinction; it can be illustrated by comparing the tetra-card games, or Good's firing squad examples, on the one hand, with the psychotherapy case on the other. In either game, the result E (a winning draw) can come about only as a result of one or the other of the causal chains specified for that game. A person who does not enter the game cannot win. In the psychotherapy example the fact that 'cures' can come about by spontaneous remission is a crucial feature. Even though we know that one or the other type of therapy (directive or nondirective) has been undertaken, we still cannot attribute the 'cure' unequivocally to the therapy.

One sort of strategy that might be suggested for the psychotherapy cases is to add another 'cause' – namely, chance. Thus, we might say, there are two initial alternatives, F (the decision to undertake psychotherapy), and another, $\bar{F}$ (to do nothing and hope for the best). Given $\bar{F}$, the probability of a 'cure' is simply the spontaneous remission rate. When $\bar{F}$ obtains, of course, neither G nor $\bar{G}$ obtains, so perhaps we should rename them G_1 and G_2 respectively (see Figure 4). It would, however, be more realistic to suppose that human beings have recuperative resources, with respect to both physical and psychological

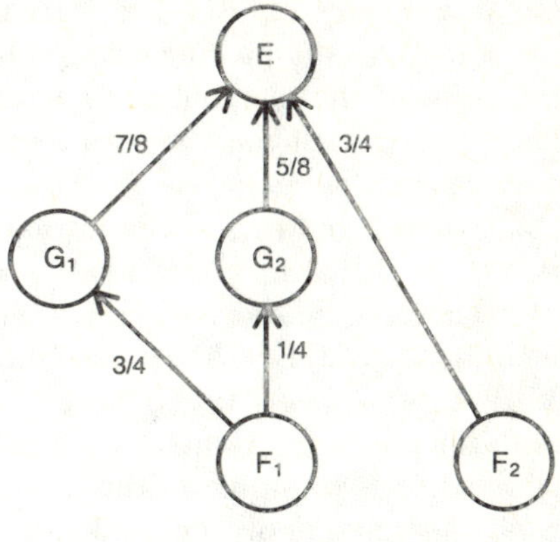

Psychotherapy - Bloggs

Figure 4

ailments. In the absence of therapy, they can sometimes produce spontaneous remissions. There is no reason to suppose that, when an individual undergoes psychotherapy, these internal resources cease operation – though the therapy might enhance or diminish them. Thus, instead of taking F and $\bar{F}$ as mutually exclusive initial states, we might designate them as F_1 and F_2, recognizing that they are not mutually exclusive. This general approach is legitimate enough, but it does not dissolve the distinction I am discussing. The fundamental distinction is between situations in which it is possible in principle to tell which of the possible causes actually brought about the effect and those in which it is impossible to tell. By observing any winning play of either tetra-card game it is easy to determine which of the two possible causes was operative in that play. By checking the number and trajectories of the bullets that entered the body of the condemned, it is possible to tell which members of the firing squad fired fatal shots. In contrast, by observing a person who underwent psychotherapy and experienced a 'cure', it is impossible in principle to tell whether the remission was spontaneous or due to the treatment. Rather, it would quite likely be more reasonable to suppose that both causes make some contribution to the remission; our problem would be to find a way to compare them quantitatively.

Consider the case of Jane Bloggs. The probability that she would get over her frigidity if she undertook psychotherapy was 13/16, while the probability that she would get over it without treatment was 3/4. She underwent psychotherapy and the problem disappeared. Did the treatment cause the remission of that symptom, or would it have vanished without psychotherapy? We could ask her whether the treatment effected the cure, but her answer should not carry much weight. The chances are that the patient wants to believe that the treatment was efficacious, and is apt to give an affirmative response as a result of wishful thinking. As I understand the situation, there is no reliable way of ascertaining which individuals who undertake psychotherapy in order to deal with frigidity (or any other problem) would have experienced remission without treatment and which would not. There is no evidential basis for asserting, counterfactually, that she would not have overcome the sexual dysfunction if she had not undergone psychotherapy, and none for asserting the contrary counterfactual. Thus, I take it, there is no reliable answer to the question, "Did the treatment bring about the cure?" All we can say is that she had treatment and her frigidity disappeared. We cannot have adequate evidence to say that she would not have been cured if she had not had psychotherapy. The case of Doakes is similar, except for the fact that there is a large discrepancy between the spontaneous remission rate and the rate of cure among those who undergo psychotherapy for problems like his. Note that when Amy Bloggs undertook nondirective therapy she reduced her chances of remission. In her case the therapy interfered with the process of recovery. Although these examples are totally fictitious, with probability values arbitrarily stipulated, I know of no reason to suppose that such interference is impossible in principle. Under these circumstances the therapy would be, in Paul Humphreys's (1981) terminology, a counteracting rather than a contributing cause.

In the cases of Jane and Amy Bloggs and Joe Doakes we have

stipulated certain probability values, and from these we can compute the Q-value – the tendency or propensity of psychotherapy to produce a cure. The Q-values differ in these two cases: for the Bloggs cousins, $Q(E':F') = 0.288$; for Doakes $Q(E:F) = 0.971$. If I understand Good's thinking on this point, he is claiming that the tendency or propensity[10] to effect a cure in either type of case is a function of the statistical relevance of the treatment to the cure. Thus, the Q-value is determined in part by the probability of spontaneous remission. This seems correct. In the case of Doakes the statistical relevance is high; in the case of Bloggs it is rather small. The tendency is, roughly speaking, the difference the treatment makes to the chance of a cure.

The problem of understanding $\chi(E:F)$ – "the degree to which F caused E" or "the contribution to the causation of E provided by F" – is rather more difficult. Intuitions apparently do simply conflict. Consider, again, the Russian roulette firing squad. It will be recalled that the command to fire received less than maximal score on the degree to which it caused the death of the condemned person. Many philosophers would make a different assignment. Those who hold a *sine qua non* conception of causation would point to the fact that, were it not for the commissar's command, the condemned person would not have died. It is true that the command is not a sufficient condition of death, but that does not make the command less of a cause when death results. Similarly, an inept assassin, who misses many of the shots fired, is totally responsible for the deaths of the victims he does kill. This contrasts sharply with the psychotherapy examples. In those cases we cannot say, but for the treatment, the problem would not have vanished. It is in cases of precisely this sort that a quantitative theory of probabilistic causality is most needed. I hope that an airing of conflicting intuitions will help us determine whether a satisfactory theory is available, and if not, help pave the way to finding one.

Department of Philosophy
University of Pittsburgh

NOTES

[1] I much appreciate the fact that Suppes (1984) gives the title "Conflicting Intuitions" to a major section of the chapter "Causality and Randomness."

[2] These values are derived as follows. For Doakes,

$$P(E|F) = 10/16; \qquad P(E|\bar{F}) = 1/100$$
$$Q(E:F) = \log\{[1 - P(E|\bar{F})]/[1 - P(E|F)]\}$$
$$= \log\{(99/100)/(6/16)\} = 0.971$$

For Bloggs,

$$P(E'|F') = 13/16; \qquad P(E'|\bar{F}') = 3/4$$
$$Q(E':F') = \log\{[1 - P(E'|\bar{F}')]/[1 - P(E'|F')]\}$$
$$= \log\{(1/4)/(3/16)\} = 0.288$$

In numerical calculations in this paper I use natural logarithms and give results to three significant figures.

[3] Another example of a causal net, which is similar in principle, but a bit more complicated, involves the occurrence of skin cancer as a result of exposure to sunlight.

[4] The relation is

$$e^{-R} + e^{-S} = 1.$$

[5] It should be recalled, in computing the value of $Q(E:G)$, that the probability $P(E|G)$ is conditioned on the state of the universe just prior to G (or $\bar{G}$), at which point the decision to undertake psychotherapy has already been made. Consequently, we need not take into account – indeed, we must not take into account – the probability of spontaneous remission in the absence of therapy.

[6] Good has kindly reminded me that in his "Errata and Corrigenda" (1962) to his "A Causal Calculus" (1961–62) he added the observation that "a 'cut' chain can often be uncut by filling it in in more detail." This *may* provide a way to handle the case of Amy Bloggs. Good's (1961–62) is reprinted, with corrections inserted, in Good (1983).

[7] Rosen's example has often been said to raise a problem concerning small probabilities. I do not think there is such a problem. In (Salmon 1980) I show that the real problem is one of negative relevance in causal chains.

[8] In (Salmon 1980, 1984) I tried to show that the phenomenon of "making it the hard way" poses very pervasive problems for some theories of probabilistic causality.

[9] Some people may be perturbed at my use of probabilities that appear to be undefined. In (Salmon 1980) I showed how to make them innocuous.

[10] It should be noted that – as Good is perfectly aware – his usage diverges sharply with that customary in philosophy. Philosophers usually use the term "propensity" to refer to a probability relation – not a statistical relevance relation – especially in the context of the so-called propensity interpretation of probability.

REFERENCES

Good, I. J. (1961–62) 'A causal calculus', *British Journal for the Philosophy of Science* **11**, pp. 305–317; **12**, pp. 43–51.

Good, I. J. (1962) 'Errata and corrigenda', *British Journal for the Philosophy of Science* **13**, p. 88.

Good, I. J. (1980) 'Some comments on probabilistic causality', *Pacific Philosophical Quarterly* **61**, pp. 301–304.

Good, I. J. (1983) *Good Thinking*. (Minneapolis: University of Minnesota Press.)

Good, I. J. (1985) 'Causal propensity: A review', in Peter D. Asquith and Philip Kitcher (eds.) *PSA 1984* (East Lansing, Mich.: Philosophy of Science Assn.).

Humphreys, Paul W. (1981) 'Aleatory explanations', *Synthese* **48**, pp. 225–232, 829–850.

Reichenbach, Hans (1956) *The Direction of Time*. (Berkeley & Los Angeles: University of California Press.)

Salmon, Wesley C. (1980) 'Probabilistic causality', *Pacific Philosophical Quarterly* **61**, pp. 50–74.

Salmon, Wesley C. (1984) *Scientific Explanation and the Causal Structure of the World*. (Princeton, N.J.: Princeton University Press.)

Suppes, Patrick (1970) *A Probabilistic Theory of Causality*. (Amsterdam: North-Holland Publishing Co.)

Suppes, Patrick (1984) *Probabilistic Metaphysics*. (Oxford: Basil Blackwell.)

I. J. GOOD

RESPONSE TO SALMON

Salmon's paper has caused me to do some rethinking that I should have done soon after reading his 1980 paper and before completing my 1980 response.

Most of my 1961 paper was concerned with causal tendency Q but there were also a few pages in which I tried valiantly to define $\chi(E:F)$, the degree to which F actually caused E, even for general networks. I am still satisfied with my explicatum for Q, and this was the topic of my lecture at this conference, but χ remains somewhat mysterious. I agree that there are conflicting intuitions about χ, and that these conflicts can occur even within one person such as myself. I think the conflicts arise partly because one can take either a reductionistic (atomistic) or a wholistic (holistic, Gestalt, Gestaltscheinlichkeitistisch, quasiconfigurationistic, syncretic, synthetic, right-brain) point of view, or maybe something intermediate. Another source of intuitional conflicts arises from the problem of deciding what the negation of B means in the notation $Q(A:B)$. I shall exemplify these conflicts in relation to Salmon's "tetracard" example after mentioning simpler examples.

General human interests in ordinary and in legal affairs are not the same as those of the "classical" (eighteenth century) physicist. Classical physics usually employs a reductionistic approach in which possibilities are broken down so to speak, whereas, outside physics, a more wholistic attitude is customary and possibilities are coalesced. (Wholistic methods do, however, occur in statistical mechanics, in Feynman's "Integration over all paths", and perhaps in most field theory.) I gave the examples of (i) the pin table, (ii) the boiling water, and (iii) the flock of birds and the exclusive club of chimpanzees. Even a single link of the firing squad, from gun to E, is really a coalescence or bundle of all the slightly different mutually exclusive paths that the bullet might take. Similarly, in the example of the pin table, all the mutually exclusive paths must be coalesced if we are to get a sensible evaluation of the extent to which the setting of the ball into motion caused its exit (event E). If we did not coalesce the (mutually exclusive) paths we would be forced to regard each path as of finite strength. Then, since

Brian Skyrms and William L. Harper (eds.), Causation, Chance, and Credence, Vol. I, 73—78.

we can, if we wish, consistently assume that the number of paths is finite, the sum of the strengths would also be finite. Yet the actual strength must be infinite, and we would run into a contradiction if we assumed the strengths to be additive.

The need to coalesce paths, and to coalesce events into a single event E (as mentioned in my lecture), in order to express the human wholistic meaning of causality, does not contradict the notion that the *probabilities* are regarded as physical (though subjectively estimated).

Consider now the network of Figure 7, where G_1 and G_2 are mutually exclusive and exhaustive, given F, and where the network has the Markov property. (The numbering of the diagrams starts with 7 to avoid clashes with the numberings in my lecture.) That is,

$$P(E|G_i \,\&\, F) = P(E|G_i) \qquad (i = 1, 2).$$

Suppose further that $P(E|G_1) = P(E|G_2)$. If we were to regard G_2 as the negation of G_1 we would deduce that $Q(E:G_1) = Q(E:G_2) = 0$ and the network would seem to have zero strength. But we clearly cannot say in general that $\chi(E:F) = 0$ under these circumstances, whether the path that obtains is $F \rightarrow G_1 \rightarrow E$ or $F \rightarrow G_2 \rightarrow E$. If we did so, we could be said to have fallen into the *Markov trap*. To avoid

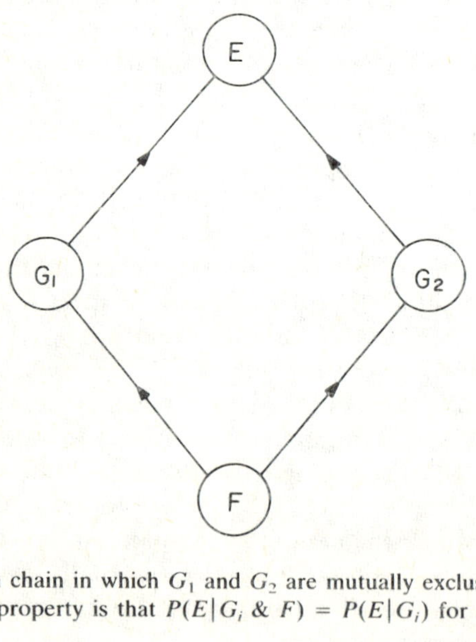

Fig. 7. A Markovian chain in which G_1 and G_2 are mutually exclusive and exhaustive. The Markov property is that $P(E|G_i \,\&\, F) = P(E|G_i)$ for $i = 1$ and 2.

this trap we must regard the negation of G_1, or of G_2, as what the state of affairs would have been if F had not occurred. This way out of the difficulty may be called *carrying along the negation* (of F).

Fortunately, in my analysis of the examples of the firing squads (with *independent* paths in which $P(E|\bar{G}_i) = 0$), and radioactive particles, the algebra comes out the same as if I had carried along the negation of F. So I believe my explicatum of Q is not undermined.

Salmon suggested that, for any network, $\chi(E:F)$ should depend directly on $P(E|F)$ as well as on the strengths of the links of the network. The notion of carrying along the negation of F, though definitely different from Salmon's suggestion, contains some of the same intuition.

But the mutually exclusive paths from F to E might have various different probabilities, $p_1, p_2, \ldots, p_n$, these probabilities being conditioned on the bundle so that $p_1 + p_2 + \cdots + p_n = 1$. If F is attained "the hard way", as in the golf example, $\chi(E:F)$ should be less than if it had been attained by a more standard route. Any causal path that occurs in ordinary life can be regarded as a bundle of nearly identical mutually exclusive paths, and we want $\chi(E:F)$ to change only negligibly if a path is split up in a niggling manner. So the strength S of a bundle should be some kind of weighted average of the individual strands. The weights can be naturally taken as the conditional probabilities of the strands, conditioned on the bundle occurring. The simplest way to achieve this desideratum is to define the strength of a path of conditional probability p_i as

$$Sp_i/\rho$$

where ρ is the average of p_i's, that is $\rho = \Sigma p_i^2$ (sometimes known as a repeat rate or as the quadratic or Gini index of homogeneity). For example, if $p_1 = p_2 = \cdots = p_n$, then the strength of any path that obtains is simply S, the same as the strength of the bundle. Again, if S is infinite, as in the model of a pin table, we obtain an infinite strength for each path, even if the probabilities of the paths are not all equal, and hence $\chi(E:F) = \infty$ whatever the path of the ball.

The situation for Salmon's tetracard example can be diagrammed as shown in Figures 8 and 9. Figure 8, which resembles Figure 7, looks like a parallel network but the paths are mutually exclusive and, in particular, are not independent given F nor given $\bar{F}$. The paths therefore have strengths (in my sense) that are not additive. But if we take a

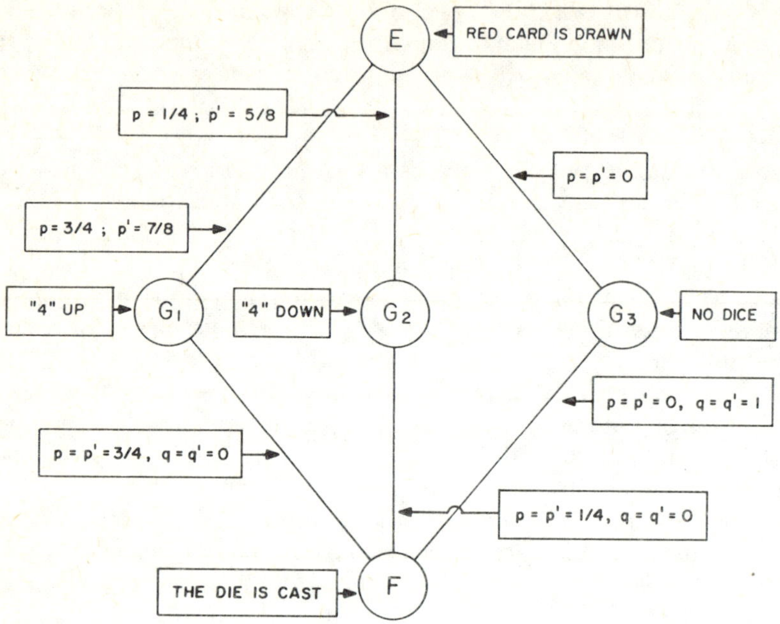

Fig. 8. *The tetracard game (Game 1).* In any link $B \to A$, P denotes $P(B|A)$ while q denotes $P(B|\bar{A})$. Moreover p' and q' refer to Game 2.

wholistic point of view, then we can lump G_1 and G_2 together and produce a causal chain as in Figure 9 where $G = G_1 \vee G_2$ and denotes the event that the die landed on the table. Similarly $\bar{G}$ denotes the event that the die remained in its box and its use in this analysis is equivalent to carrying along the negation of F. (Of course $P(\bar{G}|F) = 0$.) Figure 9 can be described as a Markov chain and its wholistic or bundled strength is log(16/6) for Game 1 if we avoid falling into the Markov trap. The probabilities of the routes $F \to G_1 \to E$ and $F \to G_2 \to E$ are respectively 9/16 and 1/16, so $p_1 = 9/10$, $p_2 = 1/10$ and $\rho = 82/100$. Thus $\chi(E:F) = (90/82)\log(16/6) = 47$ centicausats (cc) if the path is $F \to G_1 \to E$ and $\chi(E:F) = (9/82)\log(16/6) = 5$ cc otherwise. For Game 2 the corresponding values of $\chi(E:F)$ are $(546/466)\log(13/6) = 39$ cc and $(105/466)\log(13/6) = 8$ cc respectively. Thus our wholistic intuition is not contradicted. A similar analysis applies for Salmon's psychotherapy example: see Figure 10 for some of the analysis.

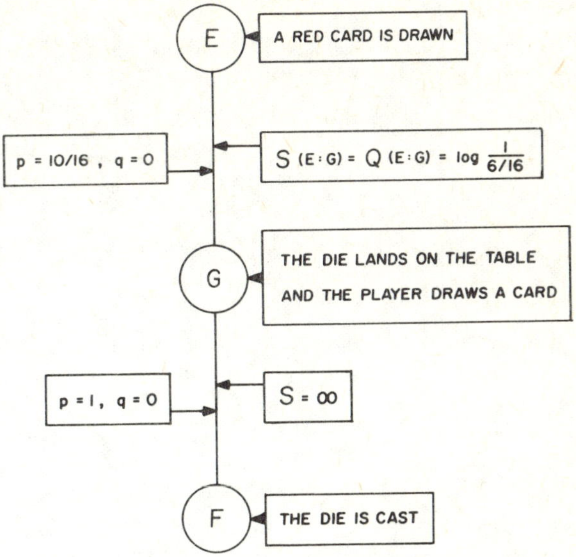

Fig. 9. *The wholistic network for Game 1.* Here G_1 and G_2 are coalesced, that is, $G = G_1 \vee G_2$.

Even if all this is accepted, there is still much to be done to define $\chi(E:F)$ to cover entirely general networks in which paths can be independent, mutually exclusive, or neither.

In Good (1980) I said I thought Salmon was confusing χ with Q, but that was because I was thinking along reductionistic lines and he was thinking wholistically. Wholistic strength is closer to Q than is reductionistic strength, whatever that may mean, and this fact misled me.

Regarding the sad case of the demise of Holmes, I have recently discovered that Watson slipped on a banana skin (left there by a chimpanzee), otherwise he would have saved Holmes's life. The main chain of causation was Moriarty–Watson–banana skin–boulder–Holmes. In my opinion it makes sense to say that Watson's act was only a weak cause of Holmes's death because of the accidental nature of the chain, and this is consistent with my present account. If a chain of causation can be broken, it can also be weakened in my opinion. To say, as Salmon did, that Watson's decision fully caused Holmes's death, seems to me to be unfair to Watson. He committed neither murder nor even manslaughter. One could reasonably claim that

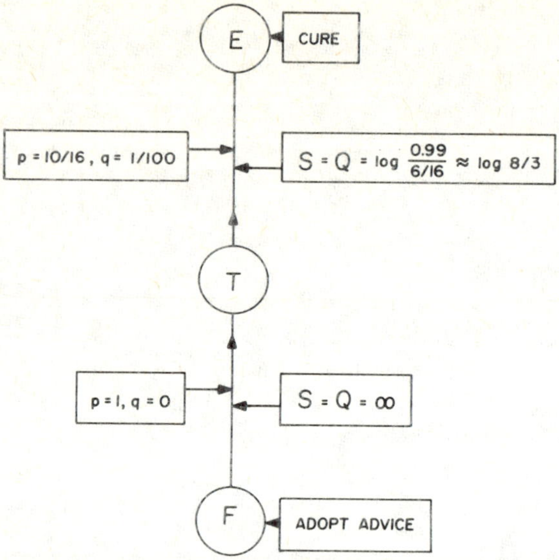

Fig. 10. *Joe's psychotherapy*. The resultant wholistic strength is approximately log(8/3). That for Jane is log(16/3).

Moriarty committed murder by provoking Watson's action. The chimp is also feeling guilty, at least that's what she told me in American Sign Language.

Salmon said we need to understand probabilistic causality qualitatively before we can produce a reliable quantitative theory. I think the attempt to sharpen either of these helps to sharpen the other. This is true in other fields such as operational research, physics, and engineering. Note that my quantitative theory for Q did begin with qualitative assumptions such as formula (7) or (7½). Also the distinction between reductionism and wholism, and the idea of "carrying along the negation" are both qualitative ideas.

REFERENCE

Salmon, Wesley C. (1985) 'Intuitions – good and not-so-good'. Previous paper in this volume.

NANCY CARTWRIGHT

REGULAR ASSOCIATIONS AND SINGULAR CAUSES

How close can we come to a Humean account of causation? Not very close at all. Hume began with cases of singular causation. He looked for something, besides spatio-temporal contiguity, to connect the cause with the effect. When he failed to find it, he moved to the generic level. This is the first thesis in the Hume program:

(1) Singular causal facts are true in virtue of generic causal facts.

But the program is far bolder than that. For at the generic level, causation is to disappear altogether and to be replaced by mere regularity:

(2) Generic causal facts are reducible to regularities.

In the past I have argued against the second of these theses: a regularity account of any particular generic causal truth – such as "Aspirins relieve headaches" – must refer to other generic causal claims if we are to pick out the *right* regularities. Hence no reduction of generic causation to regularities is possible.[1] In this paper I shall argue against the first thesis: to pick out the right regularities at the generic level requires reference not only to other *generic* causal facts, *but to singular facts as well*. So singular causal facts are not reducible to generic ones. There is at best an inevitable mixing of the two levels.

My ultimate conclusion is far stronger. Not only are singular facts not reducible to generic ones, but exactly the opposite: singular causal facts are basic. A generic claim, such as "Aspirins relieve headaches" is best reconstructed as a modalized singular claim: "An aspirin *can* relieve a headache," where the surest sign that it *can* do so is that sometimes one *does* do so.

Here I am going to focus just on the narrower claim that regularity accounts necessarily invoke singular causes, and try to show exactly why this is true. I assume that for all we know, we live in an indeterministic universe, so that if a regularity account of causation is to be given at all, it had best be a probabilistic one. I shall look at two kinds of probabilistic accounts: (a) the first – the more usual in recent philosophical literature – is based on the stratification methods used in

Brian Skyrms and William L. Harper (eds.), Causation, Chance, and Credence, Vol. I, 79–97.

social science to test causal claims; (b) the second is based on the method of the controlled experiment, typically employed in medical research. In both cases the reference to singular causes is necessary to prevent a certain kind of *averaging* that the regularity account implicitly employs. Averaging was the key to my objection to thesis (2) in the Hume program – the reason that reference to other generic claims is necessary. The averaging that affects thesis (1) is a different kind, and its solution invokes singular facts rather than generic facts. But the basic problem is the same: averaging of various kinds is bad, and pure regularity accounts necessarily average.

1. HOW SINGULAR CAUSES ENTER

The first kind of averaging creates problems because of Simpson's paradox. Let me remind you what Simpson's paradox is. The best probabilistic accounts take positive correlation as their leading idea: causes may not be constantly conjoined with their effects, but at least they should increase their frequency. Formally $P(E/C) > P(E/-C)$. But correlations can be misleading. The most widely discussed case is the problem of spurious correlations, sometimes also called the problem of joint effects. *A* may be correlated with *B* not because *A* causes *B* or *vice versa*, but rather because they are both effects of some joint cause *C*.

This is a special case of a more general problem. Here *A* and *B* are correlated with each other because they are both correlated with the third factor *C*. But the third factor need not be a cause of *A* and *B*; nor must the correlation of *A* or *B* with the third factor be positive. Any correlation with any other causal factors can render misleading correlations – or lacks of them – between *A* and *B*. This is because of Simpson's paradox: any fact about correlations in a population can be reversed by partitioning into subpopulations. Or conversely any fact true in the subpopulations can be reversed by averaging to the whole population. *A* and *B* may for instance have positive – or zero – correlation in the population, and yet be negatively correlated in both subpopulations.

The solution to Simpson's paradox is to test for correlations between *A* and *B* in very special populations – those for which all other causal factors relevant to *B* are held fixed. In such a population, there *can* be no background correlations between *C* and any other causal factor to generate misleading connections between A and B. So we look not at $P(E/C)$, $P(E/-C)$ but rather at $P(E/C \pm F_1 \ldots \pm F_n)$, $P(E/-C \pm F_1 \ldots \pm F_n)$, for each of the possible arrangements of other causal factors $F_1, \ldots, F_n$. (The partial probability that we see here is closely connected with the technique of *stratification* commonly used in the social sciences to separate out the effects of confounding variables.) Roughly then we arrive at the following formula for generic-level causal claims (See Figure 1):

$$CC \quad P(E/C \pm F_1 \ldots \pm F_n) > P(E/-C \pm F_1 \ldots \pm F_n)$$

Figure 1

I call each of these homogeneous subpopulations in which there is a fixed arrangement of the F_i *test* populations for "C causes E." Also, I will write $C \hookrightarrow E$, for "C causes E".

The formula is, as I said, rough. There are a large number of caveats and details to be filled in. There is one detail in particular that I want to concentrate on, for it is central to my claim about singular causes. Formula CC says that for a generic causal claim to hold, the cause must increase the probability of the effect in every population that is homogeneous with respect to E's other causal factors. But this condition is too strong: it holds fixed too much. We should hold fixed other causally relevant features only in individuals for whom they are not caused by C itself. The simplest example has the structure of Figure 2.

Here we have a genuine cause C, which always operates through K, K can, however, occur on its own, in which case it is still positively effective for E. If we hold K fixed, we erroneously infer that C does not cause E. For $P(E/C \pm K) = P(E/-C \pm K)$. This is a familiar point: intermediate effects in a process (here K) screen off the initial cause (C) from the final outcome (E). If we hold K fixed in this kind of case we will fail to identify genuine causes when we have them. On the other hand Simpson's paradox shows we must.

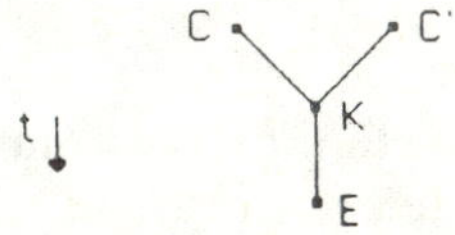

Figure 2

What are need is a more complex characterization of the test situations.

> * Test situations should be subpopulations homogeneous with respect to K, *except for those individuals where K has been produced by C*. These individuals belong in the population where they would otherwise have been *were it not for the action of C*.

In my original presentation of formula CC, I tried to avoid this problem by adding a constraint (*condition iv* in Cartwright 1983, p. 26). Elliott Sober and Ellery Eells give a concrete example endorsing the constraint:

> Surely if we accept the idea that causes increase the probability of their effects, some such condition must be right. Your phoning me causes my phone to ring; my phone's ringing causes me to lift the receiver. So, presumably your phoning me thus causes me to lift the receiver. But it shouldn't be required for this last proposition to be true that, assuming that my phone rings (or that it does not), I am more likely to lift the receiver if it is you who calls than I am if it isn't. To require this would mean that your calling at t_1 must have a way of affecting the probability of my picking up the phone at t_3 other than simply by producing the ringing at t_2.[2]

Here your phoning me = C. The ringing = K. My lifting the receiver = E. Surely, they say, "it shouldn't be required that *assuming* my phone rings – i.e. assuming K – the probability of lifting the receiver given you phone is greater than if someone else phones."[3] Given K, we no longer expect Prob $(E/C) >$ Prob $(E/-C)$; that is, we do not expect $P(E/CK) > P(E/-CK)$.

Let us call the new formula CC with the proviso * added, CC*. Here we see that singular causes have intruded. We wanted to give a pure regularity account – to give conditions for the generic claim *C causes E* – entirely in terms of conditional probabilities. But to tell *which* conditional probabilities to calculate, we must make use of singular causal facts.

2. WHY SINGULAR CAUSES CANNOT BE ELIMINATED

Prima facie then the best attempt to formulate a regularity account for generic claims requires reference to singular causal facts. But surely there must be some other way to get the right test populations without bullishly excepting certain individuals from the test population in which they appear to belong. I think not. I will tell you first how I initially began to suspect this; and second, outline the best alternative I know – one I for a while thought did solve the problem. But then I will show you how that fails.

Although CC gives an analogue to a measure of causation common in the social sciences, it was introduced into the philosophic literature about 8 years ago, primarily by Brian Skyrms[4] and by me. In my published paper I motivated the formula by a discussion of Simpson's paradox, similar to the explanation I have given here. Skyrms proceeded differently. He looked for the probabilities that were most resilient under external perturbations, and settled on the same partial conditional probabilities that I did. There is yet a third independent derivation of the formula. When I first arrived at it, I did so not in the way described today, but rather by thinking about controlled experiments. I thought I could prove the following theorem:

T: (If C is unanimous with respect to E), CC *is satisfied iff* $P(E)$ *in the test group is greater than* $P(E)$ *in the control group.*

Unanimity is a concept that we will have to think about shortly. For the moment, ignore it for it is not relevant to the issue at hand. The point here is that this theorem compares CC, and not CC* with the results of a controlled experiment. The proviso * does not enter. Nor does Skyrms in his account mention anything that looks like *. This was my ground for optimism about eliminating * altogether. But the ground was mistaken. Look back through my derivation, it turns out that * should have been there all along. The correct theorem is:

T′: (If C is unanimous with respect to E), CC* *is satisfied iff* $P(E)$ *in the test group is greater than* $P(E)$ *in the control group.*

Similarly for Skyrms' account. Essentially, Skyrms holds fixed all factors that C has no control over. But this means that the same characteristic occurring at exactly the same time must be treated differently in different individuals depending on whether it occurs in that individual on account of C, or from some other cause. The kind of condition added in * is lurking in all three derivations of the stratification formula CC. The problem of singular causes must be faced.

The best attempt to deal with the problem is due to Sober and Eells.[5] They propose holding fixed all factors prior to or simultaneous with C that either themselves directly cause (or prevent) E or that can initiate a chain of factors which can cause (or prevent) E. Their picture looks like Figure 3.

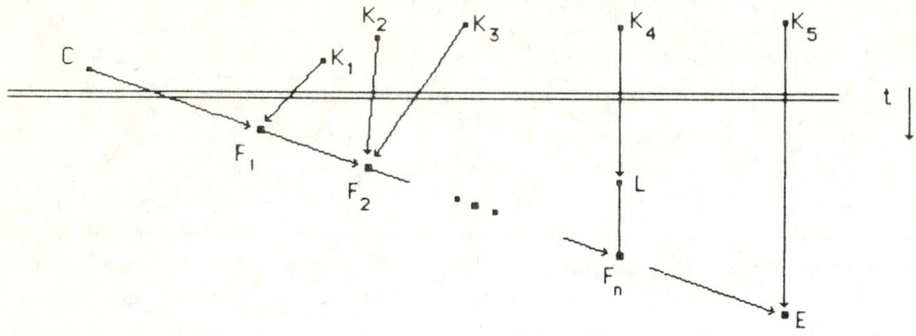

Figure 3

More formally, to test the causal law $C_{t'} \hookrightarrow E_{t''}$, Sober and Eells hold fixed all factors K_t such that $t < t'$ and (i) $K_t \hookrightarrow \pm E_{t''}$ or (ii) there exists a chain $L^1(t^1), \ldots, L^n(t^n)$ such that $K_t \hookrightarrow L^1(t^1) \hookrightarrow \cdots \hookrightarrow L^n(t^n) \hookrightarrow \pm E_{t''}$, for $t < t_1 < \cdots < t_n < t''$. I call such a test situation that satisfies (i) and (ii) *Sober and Eells homogeneous* for $C_{t'}$, $E_{t''}$.

Sober and Eells say that if we hold fixed all the *causes* of the factors, $F_1, \ldots, F_n$, we will be "paying them their due." But in fact we will not be. For we will not be considering the various combinations of these factors, one by one, as CC proposes, but averaging over them. To see the averaging, look at Figure 4. D represents a situation that is Sober–Eells homogeneous for C, E.

CC enjoins us to consider the effects of C in each of the four test situations, $\pm C' \pm P$, when produced by D. By using Eells' and Sober's

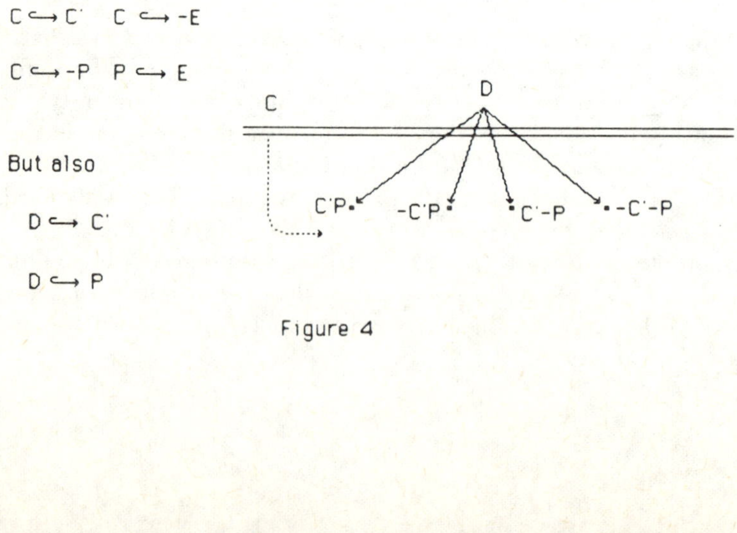

Figure 4

rule, we do not see the effects in each, but rather the effects in a population which averages over all four.

It may seem that this averaging is harmless. Recall that Simpson's paradox can arise *only when* there are correlations between C and the other relevant factors, $F_1, \ldots, F_n$. Sober and Eells assume what I call principle R, for it is essentially Reichenbach's principle of the common cause. It is also the principle that underlies the stratification procedure we see in $\mathfrak{CC}$. R says that in a population where all causes and preventatives of a factor are held fixed, the factor will exhibit no correlations with any factor that it does not cause. Thus by holding fixed all causes of our problematic K's which occur prior to or simultaneous with C, we seem to be breaking the correlations of these factors with C, *exactly when the factors are not caused by C itself.* We thus seem to avoid any problems raised by Simpson's paradox without mentioning singular causes.

But Simpson's paradox is not the whole story. There is yet another way that averaging can lead to trouble. Consider Figure 5. The numbers are a little strange because they are taken from a more detailed example that we will be studying in a moment.

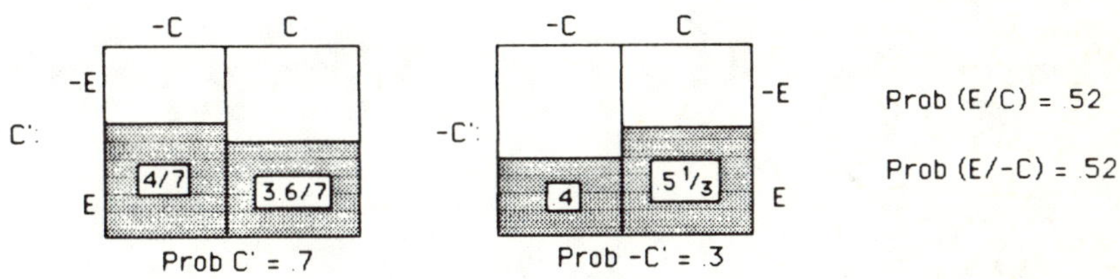

Figure 5

In $-C'$, C causes E; but in C', it prevents it. In both populations C is highly relevant. But when we average, it appears irrelevant.

You may wonder: how can C have this mixed power – sometimes to prevent E, sometimes to cause it? Here is how. Imagine that C does not affect E directly, but rather by intermediate factors, and these intermediate factors operate in opposite directions. In the language of Eells and Sober, C is not *unanimous* with respect to E.[6] (C is unanimous with respect to E if for the factors K that C causes and that themselves are relevant to E, either all K's are positively relevant or all are negatively relevant.) For example C may produce one cause of E, and yet prevent another. This kind of case has been discussed recently by Eells and Sober, by Richard Otte,[7] and by John Dupré.[8] The first illustration of it I know in the philosophical literature is by E. Hesslow.[9] I will use his example to illustrate, since it may already be familiar to many.

Hesslow argues that birth control pills may both cause and prevent thrombosis. So we let *C* equal taking birth control pills. The pills prevent thrombosis by preventing pregnancy, which is itself a cause of thrombosis. On the other hand, they may themselves cause thrombosis. Hesslow doesn't specify any intermediate steps in the positive process, but let us imagine that the pills produce a chemical, C', that causes the blood to clot when it shouldn't and produces thrombosis. Figure 6 shows the picture. (For simplicity, to make the arrows all represent "positive" causation, I have pictured: "birth control pills cause the absence of pregnancy"; rather than "birth control pills prevent pregnancy.")

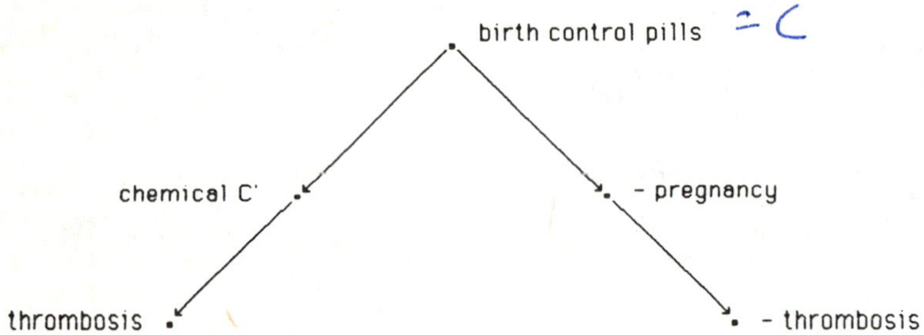

Figure 6

In his paper Hesslow considers the relative 'strengths' of the pills' tendency to produce C' and to prevent *P* and shows how these can be arranged so that the probability of an effect may *decrease*, given a cause.

But if we really want to talk about the relative strengths of C's tendencies in terms of probabilities, we must do something more complicated than Hesslow. For we know from thinking about Simpson's paradox that the probabilities must be computed in a variety of test situations. The background factors that are going to make trouble in this case are C' and *P* themselves, so we will concentrate on them. The fact that a woman already has the chemical in her blood, or will not become pregnant anyway may arise for reasons completely independent of the pills, and long before she contemplates whether to take them or not. If so, we should hold them fixed. Consider then the four homogeneous test situations; $C'P$, $-C'P$, $C'-P$, $-C'P$. Imagine the differences in the frequency of thrombosis in each test situation with the pills present and with pills absent to be given by chart 1. Chart 1′ shows how these populations change size under the action of C.

The individuals in 1.1 have the chemical condition C' so the power to produce C' plays no role. They also tend to pregnancy, and for some of these the pills will inhibit thrombosis by inhibiting pregnancy. Chart 1.2 gives the case where women neither have the chemical nor a built in safeguard against pregnancy. The increase in probability reflects the assumption that the harmful effects of the pill for thrombosis dominate its beneficial effects via pregnancy prevention. Chart 1.3 tells what happens for individuals who cannot become pregnant and do not have the chemical. Finally in chart 1.4 we see what happens for those

Chart 1

Av. P(E/C) = Av. P(E/-C) = .52

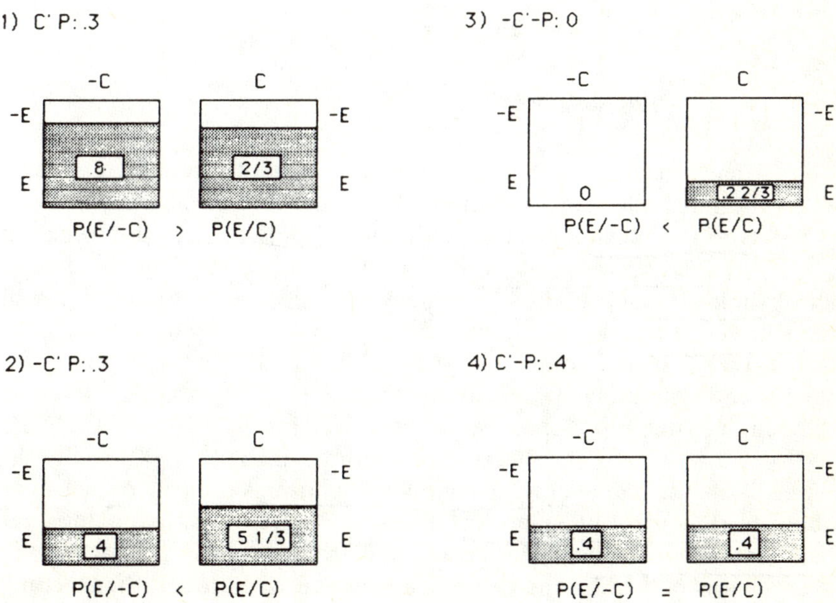

Chart 1'

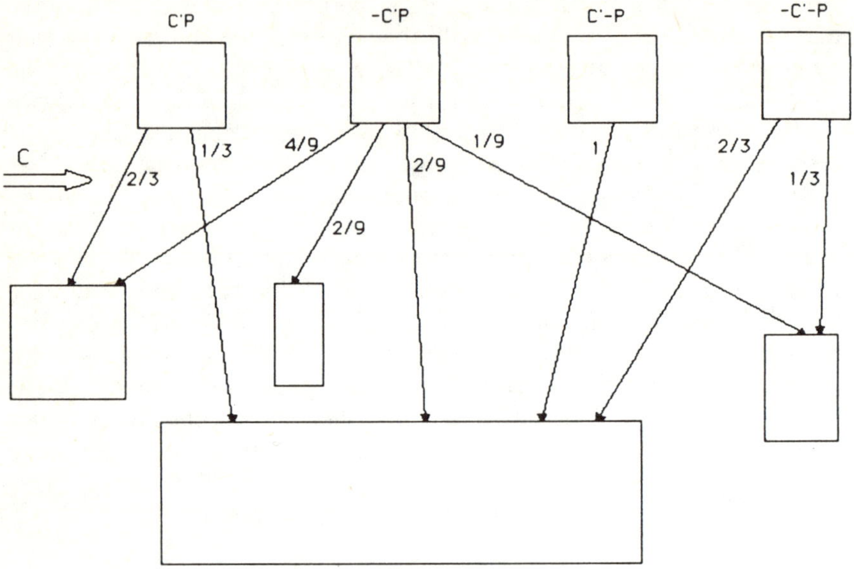

$$\varepsilon(C') = \frac{P(C'/C)-P(C'/-C)}{P(-C'/-C)} = \frac{.9-.7}{.3} = 2/3$$

$$\varepsilon(-P) = \frac{P(-P/C)-P(-P/-C)}{P(P/-C)} = \frac{.6-.4}{.6} = 1/3$$

who already have the chemical and also cannot become pregnant. Then the pills will make no difference to the chance of thrombosis.

We see in this example that birth control pills make a difference in *every single* background situation, with the exception of the one where all of their effects are already present. So they are certainly of causal relevance to thrombosis, though the relevance is mixed – sometimes they produce it and sometimes they inhibit it. But we can make this relevance completely disappear by averaging. For example, imagine the four populations appear with these weights. $C'P{:}0.3$; $-C'P{:}0.3$; $-C'-P{:}0$; $C'-P{:}0.4$. Then P(thrombosis/pills) = P(thrombosis/−pills) = 0.52. The averaging completely conceals the true causal story about birth control pills and thrombosis. The lesson is: averaging over background factors is bad. We learned from Simpson's paradox that averaging is bad when the cause is correlated with these other factors.

But here, even where such correlations are absent, we see that averaging is bad when unanimity fails.

Now let us return to Sober and Eells. The situation I have pictured is one we agree on. It is a case where the background factors, C' and P, are already in place before the time of C's operation. Nobody wants to average here – and we have seen the mistakes that can arise from doing so. But Sober and Eells' criterion implicitly performs exactly the same kind of averaging. What difference should it make whether C' and P are already in place before the decision about birth control pills is taken, or whether they befall you afterwards, so long as their doing so is completely independent of the pills? Yet in the second instance Sober and Eells average – and conclude that the pills are irrelevant to thrombosis – whereas they never average in the first.

To show this, I imagine a simple model which is described in the appendix. We consider two cases. *Case 1* is a generalization of the situation we have just looked at: Imagine we have a number of factors like C' and P operating in a Sober–Eells homogeneous background B. These factors can occur before C and if they do so, they each have a tendency to produce or prevent E. Calculate the partial conditional probability of E with and without C in every test population over these factors. This will show C's true relevance to E. It may be mixed. Now, average over these populations; that is, calculate the *total* conditional probabilities, $P(E/C)$, $P(E/-C)$. As we have seen, this can completely conceal the relevance of C. In our example, $P(E/C) = P(E/-C) = 0.52$, from Charts 1 and 1′.

Case 2. Imagine instead that these factors have a shorter causal life, so that they can act only if they occur after C. But when they do so, they act *in exactly the same way* as they do in Case 1. Sober and Eells do not consider these factors directly, but instead hold fixed only their causes. What results for $P(E/C)$ and $P(E/-C)$ when just the causes are held fixed? In our example, the appropriate distributions are those given in Chart 2. Calculating, we see $P(E/C) = P(E/-C) = 0.52$ – exactly as in Case 1. This is no accident for I can prove a general theorem, to the effect that:

THEOREM: *The Sober and Eells procedure applied in Case 2 situations gives exactly the same results as the explicit averaging of Case 1 situations.*

In sum, we agree that averaging is bad for independent factors that

Chart 2

P(E/C) = P(E/-C) = .52

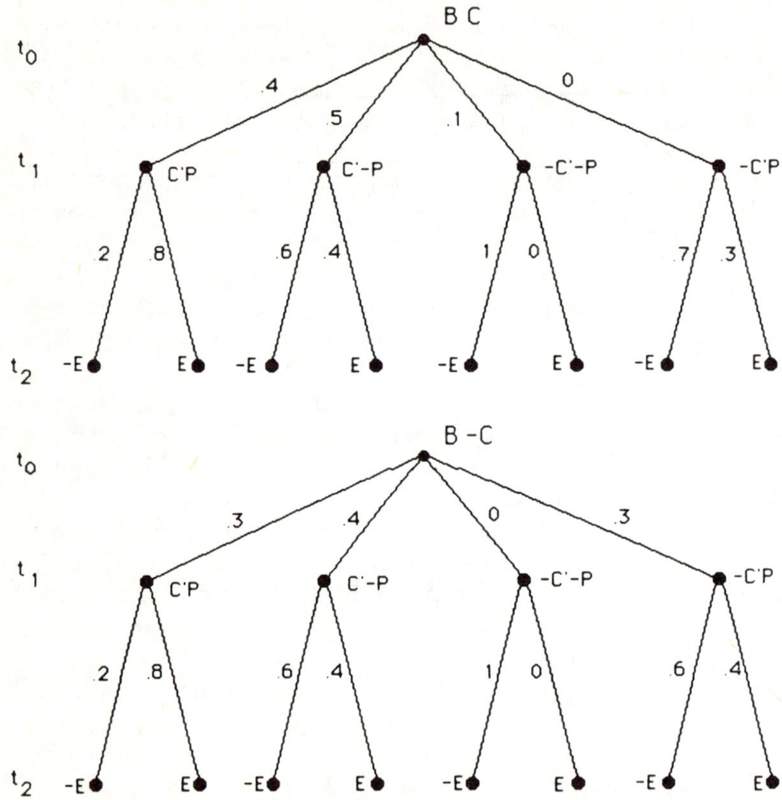

occur before C. But if these factors operate just after C rather than just before – and everything else stays the same – the Sober and Eells rule averages in exactly this bad way.

This theorem is proved in the Appendix. Here I will sketch the basic idea behind it. The theorem involves comparing two sets of numbers – those from Chart 1 with those from Chart 2. But how are Charts 1 and 2 related? Both describe what happens in a given background B that is Sober–Eells homogeneous with respect to C's causing E. I assume, as do Sober and Eells, that we are dealing here with a Markov process so that, for example, $P[E^{t+1}/(\pm C'\pm P)^t] = P[E^{t+1}/(\pm C'\pm P)^t \pm C^{t-1}]$. Chart 2 gives the distribution of C' and P at t_1 given both C and $-C$ at t_0. Some of the individuals in this distribution will have the characteristics they do because of C. Others will have been caused to do so by the independent operation of other factors in B. This is reflected in Chart 1, where C' and P are pictured as already in place when C operates.

Case 1 and Case 2 are supposed to be intimately related. All the factors are supposed to operate in *exactly the same way* in both cases, except that in Case 1, C' and P can affect E across a longer period of time; hence they must be held constant in using $\mathfrak{CC}^*$. But how do we know that? How do we ensure that the number of C' and P that C moves in Chart 1 is consistent with the numbers in Chart 2?

Recall that in both cases the background B is Sober–Eells homogeneous, so that any differences between the two graphs in Chart 2 must be due to the action of C. We ask, then, for any factor like C' or P that C causes, "Among those individuals that are $-C'$ (or $-P$) when C doesn't occur, what proportion must C affect to account for the difference between the probability of C' given C and the probability of C' without C (alternatively, P given C *versus* P without C)?" That is, we calculate $\{P(C'/C) - P(C'/-C)\}/P(-C'/C)$. This is a familiar measure. Call it *the efficacy of C towards C'*. The numbers in Charts 1 and 1′ are then computed using this measure for C's efficacy. This is the primary substantive assumption in the proof. The other assumptions are all straightforward simplifying or idealizing conditions.

There is one more topic to be taken up before concluding. I said in the introduction that I would discuss two probabilistic accounts of causation. The first is the stratification-based account that we have been considering. The second is an attempt to read a regularity analysis of causation, not from stratification methods, but from the method of controlled experiments. This is, for example, a strategy proposed by my colleague John Dupré.[10] The idea is to find an empirically acceptable description of a controlled experiment. Then, roughly, *C causes E* if and only if the probability of *E* in the treatment group is greater than in the control group.

The difficulty with this proposal is that controlled experiments average in the same way that Sober and Eells do. Other causal factors like C and P are not held fixed in a controlled experiment; rather they appear with the same distribution in the experimental groups with which they appear in the population. When we look for the probability of E at the end, we are not seeing the probability of E in each of the test groups separately as CC* requires, but rather an average over them. Because of the way C has been introduced the correlations between C and these other factors will be broken in the experiment, so problems with Simpson's paradox will not arise. But the experimental methods do not eliminate the problems that arise with averaging when unanimity fails.

Dupré does not object to this averaging. On the contrary, he endorses it. But I do not understand how he can do so. As we have seen, methods which average are not very sure methods for finding out what causes can do when they are not unanimous. Methods which average may be good guides for social decision making where we are concerned about the net effect of C across the population. From an expected utility point of view, they may even give the necessary information for personal decision making under uncertainty. That is, where we don't know which particular arrangement of externally caused factors we will fall prey to, we might resolve this problem by representing ourselves as a 'random' member of the population and averaging. But this method is no good for finding out the true causal role of C. As we have seen, C may have an effect, one way or another, on every member of the population, and yet leave the average in the population unchanged.

The conclusion is short. The best regularity accounts available fail; and once we understand how they fail, I do not see much ground for hoping we can do better. Singular causes must be brought in. This provides one very compelling reason to take seriously my more radical claim that singular causes are primary, and generic claims are just modalized versions of claims about singular causes.

APPENDIX

Consider first a 3-node first-order Markov process (it is easy to generalize to n-nodes) over C, $a = \{A_1, \ldots, A_n\}$, O, in a fixed background, B, where B is Sober–Eells-homogeneous for C, a with respect to O. Assume

(i) $\quad A_i \in a \quad \text{iff} \quad A_i(t) \rightarrow \pm O(t+1).$

(ii) $\quad A_i(t) \nrightarrow A_j(t+1),\ i \neq j; \qquad A_i(t) \nrightarrow \pm C(t+1)$

(iii) if $C \nrightarrow \pm A_i$ then

$$P[\pm A_i(t+1)/+C(t)] = P[\pm A_i(t+1)/-C(t1)].$$

This restriction comes from Sober and Eells' assumption R.

(iv) Complete independence of the A_i's under C. For any

$$B_1, \ldots, B_m \in a, m < n,$$
$$P[(B_1 \ \& \ \ldots \ \& \ B_m)(t+1)/C(t)]$$
$$= P[B_1(t+1)/C(t)] \ldots P[B_m(t+1)/C(t)].$$

From R.

The aim is to construct a 4-node process, where the probabilities will be represented by Q, in which (α) B operates entirely before C to produce exactly the same distribution over a that B by itself produces in the 3-node process. Then (β) C occurs, at T, possibly changing this distribution. Finally (γ) the members of a act exactly as they do in the 3-node process to produce O. To this end assume further

(v) $B(T-1); -C(T-1); C(T)$.

From (α), (β).

(vi) For any

$$B_1, \ldots, B_m \in a, m \leq n, Q[(B_1 \ \& \ \ldots \ \& \ B_m)(T)]$$
$$= P[(B_1 \ \& \ \ldots \ \& \ B_m)(t+1)/-C(t)].$$

From (α).

(vii) For any

$$B_1, \ldots, B_m \in a, m \leq n, Q[O(t+1)/(B_1 \ \& \ \ldots \ \& \ B_m)(t)]$$
$$- P[O(t-1)/(B_1 \ \& \ \ldots \ \& \ B_m)(t)].$$

From (γ).

(viii) For (β) we need to know: What portion of systems which are A under $-C$ would C have to affect to account for the difference in the probability of A_i when C is present from when C is absent? So define $\mathscr{E}(X)$, to be read "the efficacy of C towards X":

$$\mathscr{E}(X) = \{P[X(t+1)/C(t)] - P[X(t+1)/-C(t)]\}/ P[-X(t+1)/-C(t)]$$

Then let

$$\begin{aligned} Q[A_i(T+1)/-A_i(T)] &= 0 \quad \text{unless} \quad C(t) \to A_i(t+1) \\ &= \mathscr{E}(A_i) \quad \text{if} \quad C(t) \to A_i(t+1) \\ Q[A_i(T+1)/A_i(T)] &= 1 \quad \text{unless} \quad C(t) \to -A_i(t+1) \\ &= 1 - \mathscr{E}(-A_i) \quad \text{if} \quad C(t) \to -A_i(t+1) \end{aligned}$$

(ix) Complete conditional independence of A_i's. For any

$$\begin{aligned} &B_1, \ldots, B_m \in a,\ m \leq n,\ Q[(B_1 \,\&\, \ldots \,\&\, B_m)(t+1)/ \\ &\quad (\pm B_1 \,\&\, \ldots \,\&\, \pm B_m(t)] \\ &\quad = Q[B_1(t+1)/\pm B_1(t)] \ldots Q[B_m(t+1)/\pm B_m(t)] \end{aligned}$$

THEOREM:

$$\begin{aligned} Q[O(T+2)] &= \Sigma Q[O(T+2)/(\pm A_1 \,\&\, \ldots \,\&\, \pm A_n)(T+1)] \\ &\quad \times Q[(\pm A_1 \,\&\, \ldots \,\&\, \pm A_n)(T+1)] \\ &= P[O(T+2)/C(T)]. \end{aligned}$$

Proof.

$$\begin{aligned} &Q[O(T+2)/(\pm A_1 \,\&\, \ldots \,\&\, \pm A_n)(T+1)] \\ &\quad = P[O(T+2)/(\pm A_1 \,\&\, \ldots \,\&\, \pm A_n)(T+1)] \end{aligned}$$

by (vii). So the result will be true if $\forall m \leq n$, $B_1, \ldots, B_m$ = either A_i or $-A_i$, $A_i \in a$,

$$\begin{aligned} \Phi\colon\ &Q[(B_1 \,\&\, \ldots \,\&\, B_m)(T+1)] \\ &\quad = P[(B_1 \,\&\, \ldots \,\&\, B_m)(T+1)/C(T)]. \end{aligned}$$

Φ can be proved by induction on m. Two lemmas will be helpful. Their proof is trivial, from (viii).

LEMMA A.

$$\{1 - \mathscr{E}(\pm A_i)\}P[\mp A_i(T+1)/-C(T)] = P[\mp A_i(T+1)/C(T)]$$

LEMMA B.

$$\begin{aligned} &\mathscr{E}(\pm A_i)P[\mp A_i(T+1)/-C(T)] + P[\pm A_i(T+1)/-C(T)] \\ &\quad = P[\pm A_i(T+1)/C(T)] \end{aligned}$$

1. To show Φ for $m = 1$. By cases

(*a*) Assume $C \rightarrow A_i$. Then

$$(a_1) \qquad Q[A_i(T+1)] = Q[A_i(T+1)/A_i(T)]Q[A_i(T)] + Q[A_i(T+1)/-A_i(T)]Q[-A_i(T)]$$
$$= 1\ P[A_i(T+1)/-C(T)] + \mathscr{E}(A_i)P[-A_i(T+1)/-C(T)]$$

by vi, viii

$$= P[A_i(T+1)/C(T)],$$

by Lemma B.

$$(a_2) \qquad Q[-A_i(T+1)] = P[-A_i(T+1)/C(T)],$$

similarly using Lemma A.

$$(b) \qquad C \rightarrow -A_i.$$

Proof similar to case (a).

$$(c) \qquad -(C \rightarrow A)\ \&\ -\ (C \rightarrow -A)$$

$$(c_1) \qquad Q[A_i(T+1)] = Q[A_i(T+1)/A_i(T)]Q[A_i(T)] + Q[A_i(T+1)/-A_i(T)]Q[-A_i(T)]$$
$$= 1\ P[A_i(T+1)/-C(T)],$$

by vi, viii

$$= P[A_i(T+1)/C(T)],$$

by (iii).

(c_2) Similarly for $Q[-A_i(T+1)]$.

2. To show: if Φ is true for m, Φ is true for $m + 1$. The proof proceeds by cases.

(*a*) Consider any $B_1, \ldots, B_{m+1}$, $B_j = +A_i$ or $B_j = -A_i$, $m + 1 \leq n$, where $C \rightarrow -B_{m+1}$.

$$1. \qquad Q[(B_1\ \&\ \ldots\ \&\ B_{m+1})(T+1)] = \Sigma Q[(B_1\ \&\ \ldots\ \&\ B_{m+1})(T+1)/(\pm B_1\ \&\ \ldots\ \&\ \pm B_{m+1})(T)] \times Q[(\pm B_1\ \&\ \ldots\ \&\ \pm B_{m+1})(T)]$$

where the sum is over all arrangements of +, −.

2. $= \Sigma Q[(B_1 \,\&\, \ldots \,\&\, B_m)(T+1)]/[(\pm B_1 \,\&\, \ldots \,\&\, \pm B_m)(T)] \times Q[(\pm B_1 \,\&\, \ldots \,\&\, \pm B_m)(T)]\{Q[B_{m+1}(T+1)/B_{m+1}(T)]P[B_{m+1}(T+1)/-C(T)] + Q[B_{m+1}(T+1)/-B_{m+1}(T)]P[-B_{m+1}(T+1)/-C(T)]\}.$

By vi, iv, ix

Summing, and using viii

3. $= Q[(B_1 \,\&\, \ldots \,\&\, B_m)(T+1)][1 - \mathscr{E}(-B_{m+1})] \times P[B_{m+1}(T+1)/-C(T)]$

By Lemma A and the inductive hypothesis

4. $= P[(B_1 \,\&\, \ldots \,\&\, B_m)(T+1)/C(T)]P[B_{m+1}(T+1)/C(T)]$

5. $= P[(B_1 \,\&\, \ldots \,\&\, B_{m+1})(T+1)/C(T)],$

by iv.

For any arrangement with $-B_{m+1}$, where $C \rightarrow B_{m+1}$, the proof is similar.

(*b*) For arrangements with B_{m+1}, where $C \rightarrow B_{m+1}$, the proof is the same except in step 3, where "$[1 - \mathscr{E}(-B_{m+1})]P[B_{m+1}(T+1)/-C(T)]$" is replaced by

$$P[B_{m+1}(T+1)/-C(T)] + \mathscr{E}(B_{m+1})P[-B_{m+1}(T+1)/-C(T)] = P[B_{m+1}(T+1)/C(T)]$$

by Lemma B;
and for arrangements with $-B_{m+1}$, where $C \rightarrow -B_{m+1}$ the proof is similar.

(*c*) If $C \not\rightarrow \pm B_{m+1}$, in step 3 the replacement is by

$$P[B_{m+1}(T+1)/-C(T)]$$

in arrangements with B_{m+1}; and by

$$P[-B_{m+1}(T+1)/-C(T)],$$

for $-B_{m+1}$. By iii, this yields

$$P[B_{m+1}(T+1)/C(T)]$$

and

$$P[-B_{m+1}(T + 1)/C(T)]$$

respectively. Q.E.D.

Department of Philosophy
Stanford University

NOTES

[1] Cartwright, Nancy (1979) 'Causal laws and effective strategies', *Noûs* **13**, 419–437. Also in Cartwright, Nancy (1983) *How the Laws of Physics Lie* (Oxford: Oxford University Press).
[2] Eells, Ellery and Sober, Elliot (1983) 'Probabilistic causality and the question of transitivity', *Philosophy of Science* **50**, 35–57.
[3] Eells and Sober, pp. 39–40.
[4] Skyrms, Brian (1980) *Causal Necessity* (New Haven and London: Yale University Press).
[5] Eells and Sober, pp. 40–41. Also personal correspondence.
[6] Eells and Sober, pp. 46–47.
[7] Otte, Richard (1985) 'Probabilistic causality and Simpson's paradox', *Philosophy of Science* **52**, 110–125.
[8] Dupré, John (1984) 'Probabilistic causality emancipated', *Midwest Studies in Philosophy* **IX**, 169–175.
[9] Hesslow, Germund (1976) 'Discussion: Two notes on the probabilistic approach to causality', *Philosophy of Science* **43**, 290–292.
[10] Dupré, John (1985) 'Probabilistic causality: A reply to Ellery Eells', (Manuscript, Stanford University), which is a development of ideas in Dupré's 'Probabilistic causality emancipated.'

ELLERY EELLS

ELIMINATING SINGULAR CAUSES: REPLY TO NANCY CARTWRIGHT*

In "Regular Associations and Singular Causes," Nancy Cartwright argues that the idea of singular causation enters into, and cannot be eliminated from, the usual account of generic causal facts recently discussed by philosophers. The problem (see her article for details and references) is that it seems that the idea of singular causation must be used to decide when to hold fixed intermediate factors K, between factors C and E, when assessing the generic causal significance of C for E. Thinking in terms of relative frequency and classes, Cartwright thinks that only the subset of such a K consisting of individuals whose K-ness was *not caused* by C should be held fixed, where the other K's (those whose K-ness *was* caused by C) belong in the background context they would have actually fallen into were it not for the action of C (p. 81).

Elliott Sober and I (1983), however, have suggested a way of getting around reference to singular causal facts. The basic idea (see Cartwright's paper, or Sober's and my 1983, for details) is just to hold fixed the *causes* of K and $-K$ that are simultaneous with or prior to the time of C: by backing up like this, we claim to be paying such a K its due. Cartwright, however, objects that this involves "averaging" in a bad way, and she claims to have constructed a counterexample to the approach. I'll argue here that it's not really a counterexample.

In her example (based on an example of Germund Hesslow's 1976), C is taking birth control pills, C' is having a thrombosis-causing chemical in the blood, P is being capable of becoming pregnant, and E is having thrombosis. C causes C' and $-P$; C' causes E; and $-P$ causes $-E$. Suppose these are the only relevant factors for assessing the causal significance of C for E. The question is whether, and how, to hold the factors C' and P fixed. Cartwright considers two cases. In case 1, C' and P are thought of as factors that are present or absent already at the time of C, and are, of course, causally relevant to E. Here, I suppose, we think of C' as "chemical present at time of C" and P as "capable at time of C of becoming pregnant". Then these factors should be held fixed, and the probabilities of E given the presence and

Brian Skyrms and William L. Harper (eds.), Causation, Chance, and Credence, Vol. I, 99—104.

absence of C in the four background contexts $C'P$, $C'-P$, $-C'P$, and $-C'-P$ will show C's true causal relevance to E. In the numerical example Cartwright uses, E is, overall, probabilistically independent of C: $P(E/C) = P(E/-C)$. The latter probabilities are averages of the probabilities of E given C, and E given $-C$, within the four background contexts. Thus, averaging is bad because it does not reveal the true, *in this case mixed*, causal relevance of C for E. There is no disagreement between Cartwright and Eells/Sober about case 1.

In case 2, it is supposed that the efficacy of C' and $-P$ on E is short-lived: if C' and $-P$ are to affect whether or not E occurs, they must occur after the time of C. Since C' and $-P$ may be caused by C, and may also occur or be caused independently of C, Cartwright's approach would be to hold fixed just the subsets of C' and of $-P$ of individuals who have C' or $-P$ *independently of C*, assigning the rest to where they would have been without C. This, of course, invokes singular causal facts. Sober's and my approach is just to hold fixed the *causes* – other than C – of C' and of $-P$ that occur simultaneously with, or prior to, C. Cartwright correctly points out that we are then averaging the probabilities of E given C and those of E given $-C$ across the four contexts determined by the presence and absence of C' and of P at the later time, the time at which C' and P can have a causal bearing on E. Cartwright supposes we are in a background context B (that holds fixed the causes of C' and of $-P$) in which the values of the probabilities are the same as in the case 1 example. It turns out again that $P(E/BC) = P(E/B-C)$.

As Cartwright points out, nobody wants to average in case 1. Sober and I average in case 2. But Cartwright asks, "What difference should it make whether C' and P are already in place before the decision about birth control pills is taken, or whether they befall you afterwards, so long as their doing so is completely independent of the pills?" (p. 89); and she says, "Yet in the second instance Sober and Eells average – and conclude that the pills are irrelevant to thrombosis – whereas they never average in the first" (*ibid.*).

I think it is incorrect to average in case 1, yet correct in case 2. To show this for case 2, it is necessary to look more closely at Cartwright's example. In the absence of C, the factors C' and P are distributed among the members of the background set (or context) B as shown in the top row of the chart below. The presence of C causes shifts in this distribution, as indicated by the arrows (for example, 2/3 of those in

$C'P$ remain there and 1/3 go to $C'-P$). (The ratios given in my chart differ in some cases from those in Cartwright's Chart 1′, but I think those given below are correct.) The bottom row of boxes indicates how C' and P distribute themselves among the members of B when C is present. And below the boxes, I have indicated the probability of E in the corresponding sets (or contexts); for example, $P(E/C'P) = 0.8$. Assuming that all chains are Markovian (i.e., that whether or not C is present does not affect the probability of E once the presence or absence of C' and of P is specified), the chart gives all the information necessary for calculating $P(E/BC)$ and $P(E/B-C)$. It turns out that $P(E/BC) = P(E/B-C) = 0.52$.

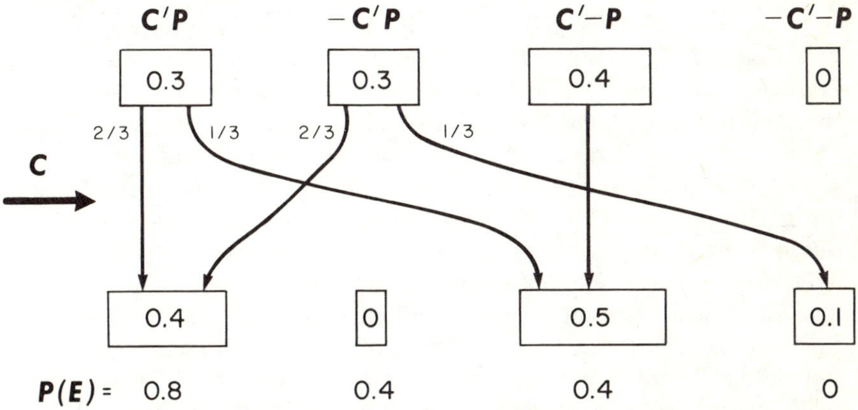

It must be kept in mind that the probability values given in the chart pertain to a particular background context B in which all factors, other than C, that are causally relevant to C' or to P are held fixed in one way. This particular constellation of background factors confers the probabilities on the top row to $C'P$, $-C'P$, $C'-P$, and $-C'-P$ when C is absent; and it confers the probabilities on the bottom row to these four factors when C is present. Also, it is obviously an indeterministic matter in this example which (if any) individuals in B will get C' and which (if any) will have P. For if this were a deterministic situation, then holding fixed (in B) all factors except for C that are causally relevant to C' or P, and then adding C or $-C$, would make for all 0–1 probabilities in the top two rows. B together with whether or not C are all that is causally relevant, at the time of C or $-C$, to whether or not one will get C' or get P.

So, given that all chains are Markovian, as Cartwright assumes, B, together with whether or not C, are all that is causally relevant, at the time of C or $-C$, to whether or not one will get E. *BC and $B-C$ specify (or hold fixed in one way) everything that, at the time of C or $-C$, is causally relevant to whether or not E will occur*. Thus, since BC and $B-C$ confer exactly the same probability on E, it is evident that C and $-C$ *make no causal difference, in background context B, as to whether or not one gets E*.

Cartwright has the intuition that "averaging" and looking at just $P(E/BC)$ and $P(E/B-C)$ conceals part of the true causal roles of C and $-C$ for E in context B. My intuition is that C and $-C$ make no difference for E in B, as the equality of the last mentioned probabilities indicates. Perhaps the source of Cartwright's intuition is that C *does* make a difference in *the probabilities of various "ways" in which E might come about, in B*. For example, C makes it more probable that you will get E "via" (or perhaps I should just say "in the presence of") $C'P$ than $-C$ does; and C makes it less probable that you will get E "via" (or in the presence of) $-C'P$ than $-C$ does. But this is quite distinct from C's making a causal difference in simply *whether or not one will get* E, in B.

Here is a simpler example than Cartwright's, but one that I think is analogous in all relevant respects, and one that I think clearly illustrates the distinction just mentioned. I am going to reach into an urn and pull out a ball. There is now an equal number of red, green, and yellow balls. After I pull out a ball, I'll role a fair die. If I draw a red ball, you get \$1 if the die lands ace, nothing otherwise; if the ball I draw is green, you get \$1 if the die lands with an even number up, nothing otherwise; and if the ball is yellow, you get \$1 if the die lands with a 2–6 up, nothing if it lands ace. Now (before I draw from the urn) you have a choice. You may paint all the red and yellow balls green; this is C. Or you may leave the balls in the urn alone; this is $-C$. Factors R, G, and Y are my drawing a red, a green, and a yellow ball, respectively. And E is your winning the \$1. The background context can be thought of as given by the original constitution of the urn, and perhaps other features of the situation as described above. The situation is represented by the trees below, which are similar to Cartwright's Chart 2 for the birth control example.

Now the question is: What is the causal role of C for E? It's easy to see that $P(E/C) = P(E/-C) = 0.5$. I think it is clear that in this

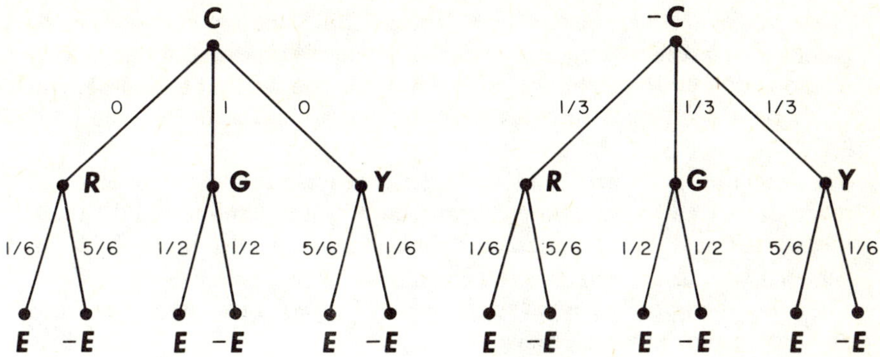

example we have to say that C is causally neutral for E. Although C has a causal effect on *what color ball is drawn* – on what "ways" are potentially available for winning the \$1 – it doesn't causally affect *the chance of E's occurring*.

On Cartwright's approach, however, it turns out that there is no single causal role of C for E in this example. Paraphrasing her characterization (*, p. 82) of appropriate background contexts (or test situations), and adapting it to this example, we have:

> Test situations should be subpopulations homogeneous with respect to (the intermediate causal factor of) color, except for those individuals whose color has been produced by C. These individuals belong in the subpopulation where they would otherwise have been were it not for the action of C.

In other words, we should in this example hold fixed *the original color of the drawn ball*. Then for individuals for whom a ball is drawn that was originally red, C is causally positive for E; for individuals for whom an originally green ball was drawn, C is neutral for E; and for individuals for whom the drawn ball was originally yellow, C is negative for E. It seems inappropriate, however, to hold fixed the original color of a drawn ball, for there is nothing about a given individual – you for instance in our example – *at the time of C or −C* in virtue of which, say, an originally red ball will be drawn; which ball will be drawn (I am assuming for the example) *is the outcome of a chance process*. And we're interested in the causal significance of C *at the time of its occurrence* for E at the later time.

For perhaps on even clearer case, suppose C is the act of painting all of the red and green balls yellow, and painting one of the yellow balls green. Surely this is a case in which *C* is causally positive for *E*, and *C* would surely be the right option to choose if you wanted to win $1. But Cartwright's approach implies that there is no single causal role of *C* for *E* in this case either.

I conclude that Cartwright's alleged counterexample to Sober's and my approach to intermediate causal factors is not genuine. And I think the appearance of a counterexample is based on mistaking *C's being generically causally relevant to the potential proximate causes of E* for *C's being generically causally relevant to E itself*, in a case where this distinction makes a difference.

University of Wisconsin-Madison

NOTE

* I thank the John Simon Guggenheim Memorial Foundation for financial support during the time this paper was written.

REFERENCES

Cartwright, N. (1987) 'Regular associations and singular causes', previous article in this volume.

Eells, E. and Sober, E. (1983) 'Probabilistic causality and the question of transitivity', *Philosophy of Science* **50**, 35–57.

Hesslow, G. (1976) 'Two notes on the probabilistic approach to causality', *Philosophy of Science* **43**, 290–292.

NANCY CARTWRIGHT

REPLY TO ELLERY EELLS

Not surprisingly, I do not agree with Ellery Eells that in his example "*C* is causally neutral for *E*." I do agree that *C* "doesn't affect *the chance of E's occurring*," nevertheless I think that it is a true causal "law" in Eells' constructed universe that *C*'s cause *E*'s. It is also a true law that *C*'s prevent *E*'s. His example resembles in this respect the version of Hesslow's example in which the contraceptives both cause and prevent thrombosis although, all-told, they leave the probabilities unaffected.

To argue for this, I shall change the content of the example yet again, since I find that causal intuitions about games of chance are generally unclear. Imagine instead that the probabilities Eells describes arise in a ruthless experiment designed to test three new drugs that may be used to treat manic-depressives. The subjects will randomly be given red, green, or yellow pills, with equal probability. Those who get the yellow pills have a 5/6 chance of becoming so depressed that they commit suicide; those given red pills have only a 1/6 chance; with green the chances are 50–50. Now you have the option to substitute green pills for the red and yellow ones. Surely this action has a significant impact on the survival or death of the subjects, even though it leaves their relative numbers unchanged.

Perhaps the difference in intuitions between Eells and me comes from the difference problems that motivate our studies of probabilistic causation. I have been concerned with the metaphysical question, What is a causal law; Eells has been concerned with questions of strategy. The two are connected I think, but not in such a simple way that you can read the causal laws that govern individuals from the strategies they should choose. I mention this difference because it seems relevant to Eells' "clearer case" where the option is to paint all the red and green balls yellow and one of the yellow balls green. As Eells says, the formula I propose implies that this action does not have a univocal causal role. Unlike Eells I think this is the correct thing to say. This action, by causing a number of balls to be yellow, causes winning; by preventing one from being yellow, it also prevents win-

Brian Skyrms and William L. Harper (eds.), Causation, Chance, and Credence, Vol. I, 105–108.

ning. We both agree that it "would surely be the right option to choose if you wanted to win $1," but I do not think that this bears on its univocality. Here is a single action with opposed causal tendencies. It is the right one to choose because the positive tendencies outweigh the negative.

Turning explicitly to questions of metaphysics I take it that the causal law *C's cause E's* tells us in part that *C*'s, just by being *C*'s, can produce *E*'s; and I suppose that this blatantly non-Humean fact will manifest itself in other states of affairs that are more empirically accessible, in observable correlations, for example. Philosophical studies that go under the name "probabilistic theories of causality" can tell us exactly which observable correlations will serve as reliable evidence for the causal law. When we focus on this question of what causal laws are true, I think we can see clearly why we should not allow the action in Eells' first example to be causally neutral. If you substitute green pills for red, then on average out of every hundred subjects, 22 will die who would not otherwise, and they do so just on account of your action. That seems to be conclusive evidence that this action can cause death. In this experimental set-up, the "law" *C causes death* is true. Since the *C* in question also involves substituting green pills for yellow, it will in addition be the case that out of every hundred subjects there will be, on average, 22 different individuals who survive who wouldn't have done so otherwise. So it is also true that *C* prevents death. This may leave the moral status of *C* unclear; but it shows that *C* is in no way causally neutral.

Eells says he does not like the outcome of my account because it does not reflect the causal significance of *C* at the time of its occurrence for *E* at the later time. But I do not see why this particular *C* is in any worse state than any purely probabilistic cause. *C* has the power at *t* to produce *E* later. I agree that whether it succeeds in doing so in any particular case depends on no further facts true at *t*. But that is always the case where causes may be indeterministic.

I think Eells' worries about the time have a different source, which can be seen in the first half of his contribution to this volume. It may be that he is unhappy with the idea that two contrary causal laws can both be true together, simpliciter. In "Probabilistic Causal Levels", section 2, Eells proposes relativizing causal laws to populations. Once this is done, a causal law will be true in a population just in case the cause increases the probability of the effect in that population. This means that we will never get contrary causal laws, since we cannot have both $P(E/C) > P(E/-C)$ and $P(-E/C) > P(-E/-C)$ in the same population. In the Hesslow example, for instance, Eells says that contraceptives are causally positive for thrombosis "in the subpopulation of women who become pregnant (a few despite taking oral contraceptives)." They are also causally positive "in the subpopulation of women who do not become pregnant (for whatever reason)." In both these cases the probability of thrombosis with the pills is greater than without. In the version of the example where the reverse is true in the total population of women, Eells concludes, "in the whole population, C (contraceptive-taking) is negative for E (thrombosis)."

I have been reluctant to adopt this proposal for a couple of different reasons. First, I don't know quite how to read Eells' "in the population . . .". For the whole population is just the union of the two subpopulations he delineates, yet different causal laws are true in the population, taken as a whole, than taken as the union of its two parts. It would seem less counterintuitive if we were relativizing to descriptions rather than populations. But that kind of manoeuver tends to lead to a more subjectivist view of causal laws than, I think, either Eells or I want; that is, what causal laws are true comes to depend on one's choice of description. Possibly here we just need to learn more about how Eells intends his relativization program to go.

Second, under any natural reading, it seems to me false that in the subpopulation of women who do not become pregnant, contraceptives are only positive for thrombosis. It might well be the case that in this very population most of the women have been saved from thrombosis by the pills' action in preventing their pregnancy. I look at this population differently from Eells. It is not a special subpopulation in which different causal laws are true than in the total population; it is rather a special subpopulation in which it is appropriate to test one of the causal laws that may be true in the whole population.

Eells and I share the view, contrary to Richard Otte for instance, that causes increase the probability of their effects. But I take this to mean that in the appropriately selected test population the probabilities should increase. It doesn't mean that the laws are true only relative to the populations in which the probabilities go up. I think that what Eells is looking at in each case is not the full set of causal laws true for a population, but rather what John Dupré and I have called "the net causal upshot", of the putative cause in that population.*

I would like to make two minor remarks before closing. The first is to point out a way in which Eells' and my example differ. Mine is more complicated because I consider intermediate factors – the chemical in the blood and pregnancy – that are produced independently of each other; whereas drawing a red, yellow, or green ball in Eells' example are incompatible events. I don't know if this makes a difference, but it would be a place to look for those who turn out to share my view about my example and Eells' view about his example.

The last remark is just to concur with Eells that there does seem to be in many cases a distinction between generically causing a proximate cause of E, and generically causing E. But it will be a hard one to draw since most legitimate generic-level causes operate by producing more proximate causes. My suspicion is that getting the distinction straight will require richer concepts than those allowed in probabilistic theories; but perhaps this is an excessively pessimistic view.

Stanford University

NOTE

* Dupré, John and Cartwright, Nancy (1986) 'Probabilistic capacities and probabilistic laws'. Manuscript (Stanford University: Stanford, Cal.).

ELLERY EELLS

PROBABILISTIC CAUSAL LEVELS*

1. INTRODUCTION

According to a standard theory of probabilistic causality, causes C raise the probability of their effects E within causally homogeneous background contexts. More precisely, let K_i's be maximal conjunctions of factors – not including C, not-C, or effects thereof – that are causally relevant to E. Then C causes E if and only if, for each i, $Pr(E/K_i\&C) > Pr(E/K_i\&{}^{\sim}C)$.[1] Alternatively, we may call this relation that of C's being *causally positive* for E, and characterizations of the relations of C's being *causally negative* and *causally neutral* for E can be obtained by substituting "<" and "=", respectively, for ">". Of course, the role of the idea of background contexts in the theory is to avoid getting the wrong answer in cases of the well known phenomenon of "spurious correlation", which arises in cases of Simpson's paradox, for example. According to the Pareto-dominance revision of the theory suggested by Skyrms (1980), C need only raise the probability of E within *some* background context K_i, and must not lower the probability of E in any. Although this revision is perhaps more plausible, I will confine my attention to the "always raises" version; the issues relevant to which of the two versions is more plausible are irrelevant to the issues I wish to discuss, and this will simplify my discussion.

The condition that a cause must "always" raise (or, for the last time, for the Pareto-dominance revision, "never" lower but "sometimes" raise) the probability of its effects has received considerable attention. Actually, there are two ways in which one can take the question, "Must a cause *always* raise the probability of an effect of it?", depending on how you read "always". On the one hand, one may ask, "Must a cause really raise the probability of its effect *in every background context*?" Alternatively, one may ask, "Must a cause *in every particular actual joint instance of it and an effect of it* raise the probability of its actual effect?" Note that this second question is not the same as, "Must every particular instance of a cause raise the probability that an

Brian Skyrms and William L. Harper (eds.), Causation, Chance, and Credence, Vol. I, 109–133.

instance of an effect of it will occur?", for this, given sufficiently fine grained background contexts, is equivalent to the first question. The second question pertains to actual joint occurrences of a cause and effect; it presupposes that the cause and effect both actually happen, whereas the first question does not. The second question can also be put thus: "If an occurrence of *E* is followed by an occurrence of *C*, must *C* have raised the probability of *E* in order for it to qualify as a cause of *E*?"

Hesslow (1976), Dupré (1984), and Otte (1985) have argued that a genuine cause needn't raise the probability of its effect in every background context. For example, Dupré has suggested that if scientists discovered a rare physiological condition under which smoking actually decreased the probability of lung cancer, we would not say that they had thereby discovered that smoking does not after all cause lung cancer (p. 172). And Hesslow and Otte discuss the following somewhat different kind of case. Taking oral contraceptives causes thrombosis, but, since it also prevents pregnancy, and since pregnancy causes thrombosis, taking oral contraceptives also prevents thrombosis. If the negative effect of the contraceptives on pregnancy, and nonpregnancy's negative effect on thrombosis, are both strong enough, then, overall, taking oral contraceptives will lower the probability of thrombosis – even though, "by hypothesis", taking oral contraceptives is a cause of thrombosis. Hesslow concludes: "*it is entirely possible that a cause should lower the probability of its effect*" (p. 291, his emphasis). Otte agrees, but adds that taking oral contraceptives also prevents thrombosis, so that some causes have a "dual nature".

I will discuss these examples, and the arguments given by Dupré, Hesslow, and Otte for a negative answer to our first question, below. The following examples are intended to clarify our second question, and illustrate some of the considerations that have been taken to be relevant to what our answer to it should be. To a certain extent (and definitely in the case of I. J. Good), such examples have been invented to cast doubt on the idea that causes must raise the probability of their effects in the sense of the second question, thus pointing to a negative answer.

Consider this example of Deborah Rosen's (1978, see also Suppes 1970). A golfer makes a shot, the ball hits the limb of a tree, and then, improbably enough, the ball is deflected directly into the cup for a birdie. It is supposed that the probability of a birdie is low given the shot in the first place, but that when you add the information that the ball will hit the tree, the probability of a birdie becomes even lower. So, even though the ball's hitting the tree is what *actually caused* the birdie, its hitting the tree nevertheless *lowers the probability* of the birdie. (Several reactions to, and clarifications of, this example will be discussed below.) I. J. Good (1961–2, 1983, 1985) has given the following example. Sherlock Holmes is at the bottom of a cliff; at the top are Watson, Moriarty, and a loose boulder. Watson knows that if he does nothing, Moriarty will push the boulder off the cliff in just the right way to kill Holmes. The only way to save Homes is for Watson himself to push the boulder off, trying to push it hard enough to overshoot Holmes. Watson tries, but the boulder hits and crushes Holmes anyway. Again, although Watson's push is what *actually caused* Holmes' to be crushed by the boulder, his push nevertheless *lowers the probability* of that effect.

Good concludes from this example that there are two distinct causal concepts, "the tendency of F to cause E and the degree to which F actually caused E" (1985, p. 831), and he offers analyses of each concept. Elliott Sober (1985) agrees with the two-causal-concept idea, and, after discussing an example similar to Rosen's (see below), concludes that the probability increase model of causation is true only at the level of *property causation* (the "tendency to cause" level of our first question above), and not at the level of *token causation* (the "actually caused" level of our second question).

I believe that, at the level of properties, we should require that a cause "always" (i.e., in every background context) raises the probability of its effect. My reasons for thinking so, despite the examples of Hesslow, Dupré, and Otte, will be given in the next section. Following that, I will consider the question of probability increase at the level of token causation. There, I will argue for a basically three-way classification scheme of the ways in which one token event can be token-causally significant for another token event. The result is a conception of singular causation that is closer to the probability increase model than might initially seem plausible in light of examples like Good's and Rosen's. The discussion in that section will be somewhat tentative, speculative, and sketchy; the theory developed will rely on intuitions relating to a pair of examples. In the final section, I will consider the reactions of Rosen, Good, and, briefly, Salmon, to cases of token causation of the kind in question.

2. UNANIMITY

Following John Dupré, let us call the condition that a cause must raise the probability of its effect in every background context the condition of "contextual unanimity". I believe that the temptation to doubt this condition when confronted with examples like Dupré's results from not fully appreciating the following feature of the standard theory of probabilistic causality outlined above: according to the theory, probabilistic causality is actually a relation among *three* things, a causal factor *C*, a probabilistic effect *E* of it, and *a population within which C is a cause of E*. I doubt very much that, except in cases of deterministic causation, a cause can be contextually unanimous for an effect of it in every population no matter how broad. No doubt we can imagine a kind of individual for whom smoking would lower the probability of lung cancer: perhaps a person who would be forced to run (not walk) a mile for every cigarette smoked – or a person for whom smoking's effect on the heart strongly dominates its effect on the lungs, so that the person, if a smoker, would probably die of a heart attack before having a chance to develop lung cancer.

The correct approach to such cases is, I believe, to say that *in the subpopulation of individuals who lack the relevant rare condition*, smoking is causally positive for lung cancer; *in the small subpopulation of individuals who have the condition*, smoking is negative for lung cancer; and in the combined population, there is no single causal role of smoking for lung cancer. This approach, exploiting the relativity of probabilistic causality to a population, saves contextual unanimity and at the same time captures all of the causal truth.

In any case, there is a problem for those who would suggest that the condition of contextual unanimity be abandoned: how small does the small subpopulation in which the causal role of *C* for *E* is reversed have to be (or how significantly must the causal role of *C* for *E* in that subpopulation be reversed) for us not to require *C*'s raising the probability of *E* in it? Consider this hypothetical example. Suppose that smoking sharply increases the probability of a heart attack for non-exercisers, but that for exercisers, smoking equally sharply lowers the probability of an attack. If we require contextual unanimity, then smoking is neither positive nor negative (nor neutral) for heart attacks in the population as a whole (assuming that there are both exercisers and nonexercisers in the population). But if we abandon contextual unanimity, then whether smoking is causally positive or causally negative for heart attacks in the population will turn on the population frequency of exercisers. This seems wrong.[2] Indeed, a person contemplating smoking, and assessing the health risks, should be concerned with whether or not he is an exerciser, and not with what the population frequency of exercisers is. For the causal truth is: smoking causes heart attacks in the (sub)population of nonexercisers and prevents heart attacks in the (sub)population of exercisers, and smoking has no single causal role in the whole population (positive, negative, or neutral).

Now consider the example Hesslow and Otte discuss. In this case, it is not *contextual* unanimity that is in question, but rather what Dupré has called "unanimity of *intermediaries*". In this example, taking oral contraceptives affects the chances of thrombosis by way of two causal paths. On the one hand, taking oral contraceptives (C) prevents pregnancy (F) and nonpregnancy prevents thrombosis (E). So, taking only this into account, it seems that taking oral contraceptives prevents thrombosis. However, by way of a different causal path, taking oral contraceptives causes thrombosis; say the former causes factor G and G causes thrombosis. (See Figure 1.) Unanimity of intermediaries is the requirement (should one wish to adopt it) that a cause raises the probability only of intermediate factors that are positive for an effect of it. In this example, unanimity of intermediaries fails. If we *reject* unanimity of intermediaries, then whether C is positive or negative for

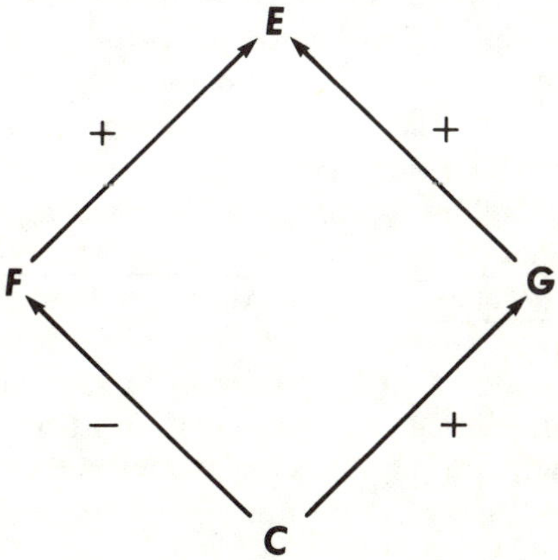

Fig. 1.

E will depend on the relative "strengths" of the various causal chains from C to E. In the example, if C is sufficiently negative for F and F is sufficiently positive for E (where what is sufficient depends on the degree to which C is positive for G and the degree to which G is positive for E), then C will be causally negative for E; otherwise, C may be positive for E. If we *adopt* unanimity of intermediaries, then we must say that there is, in some sense, no single causal role of C for E, that the cause has, as Otte puts it, a "dual nature".

Elliott Sober and I (1983) have pointed out that unanimity of intermediaries is not a part of the standard theory of probabilistic causality sketched above, and we have shown – essentially by pointing out again the kind of possibility realized in Heslow's example – that unanimity is not a necessary condition for transitivity of causal chains, though sufficient when combined with the Markov condition and a natural independence condition. What are we to make of this in light of Heslow's and Otte's contention that taking oral contraceptives is, in the example, causally positive for thrombosis, even if the effect of the contraceptive on pregnancy is sufficiently negative, and the effect of nonpregnancy on thrombosis is sufficiently negative, that taking oral contraceptives lowers the probability of thrombosis – indeed, *within every causally homogeneous background context*?

I suggest that sensitivity to the fact that the standard theory of probabilistic causality is a three-place relation – one place filled by a population – will again point to the correct approach. I suggest that, relative to the whole population (of women), taking oral contraceptives is (in the example) causally negative for thrombosis, as the following remarks of Otte himself strongly suggest: "if C occurs and everything else remains fixed, then the probability of E occurring is lowered, and if C does not occur and everything else is fixed, the probability of E is higher" (1985, p. 122). Otte rejects this argument, saying that it fallaciously supposes that a factor cannot be both positive and negative for a second factor. Although I don't see the force of Otte's objection as against the argument he considers, I do have sympathy with the idea that a cause can have a "dual nature" (i.e., that unanimity of intermediaries can fail), and I believe that the probabilistic theory adequately reflects the "dual nature" of many causes, as the following suggestions pertaining to the example under consideration would seem to bear out.

In the subpopulation of women who become pregnant (a few despite taking oral contraceptives), taking oral contraceptives increases the probability of thrombosis; and also in the subpopulation of women who do not become pregnant (for whatever reason), taking oral contraceptives increases the probability of thrombosis. In these two subpopulations, *C* is positive for *E*; even though, as we have seen, in the whole population *C* is negative for *E*.[3] Note, incidentally, that this approach does not run afoul of Cartwright's (1979) requirement that we should not hold fixed, in background contexts, factors causally intermediate between *C* and *E* when assessing *C*'s causal impact on *E* in a population. There is a difference between, on the one hand, investigating causal relations in the subpopulation of *F*'s and, on the other hand, holding *F* fixed in a larger population. And different considerations (both pragmatic and epistemic) are relevant to the "appropriateness" of investigating *F*'s separately, on the one hand, and holding *F* fixed in a larger population, on the other.

I conclude that the right answer to our first question is "yes": a causal factor in a population should be required to raise the probability of its effect *in every background context*. In needn't however be causally positive only for intermediate factors that are themselves positive for its effect. By requiring contextual unanimity, rejecting unanimity of intermediaries, and exploiting the fact that probabilistic causality is a relation between three things, including a population, the standard theory appropriately accommodates all the property level causal truth, without allowing us to go beyond what the causal facts allow.

3. TOKEN CAUSATION

Let us add the following instructive example to the cases given above in which a token cause lowers the probability of its token effect. It is a modification of the Rosen example, a version of which has been discussed previously in Eells and Sober (1983) and in Sober (1985). The golfer swings, the ball rolls straight towards the cup, and a squirrel comes along and kicks the ball. Kicks of exactly the same kind as this kick have a tendency to prevent balls rolling just as this one was from going into the cup. It is, let us assume, an irreducibly statistical fact about just this kind of situation, with the ball rolling just as it was and the kick administered just as it was, that the probability of a birdie is lowered from 0.98 to 0.1 at the time of the kick. In this particular case, however, the ball headed straight for the cup after the kick and fell in for a birdie. I assume everyone would agree that the kick *caused* the birdie, even though kicks just like this one *lower the probability* of birdies in exactly the kind of case described.

To sharpen our intuitions, we may assume that the squirrel kick resulted in the ball's taking a quite different path to the cup than it would have without the kick. This is possible because the course is sloped, allowing for different paths to the cup. And all this is consistent with the assumption that, if the situation were replicated 1000 times, there would be only about 100 birdies: the probability of a birdie given the squirrel kick (0.1) is lower than the probability of a birdie given no kick (0.98). Of course, the example assumes physical indeterminism.

In this example, as well as those of Good's and Rosen's, there is no denying that the token event that *actually caused* the relevant token effect is of a kind that is a *negative causal factor* (in the sense of the previous section) for the kind instantiated by the effect. And each of these examples can be elaborated in such a way that no matter how much physical detail we include in the description of the token cause, that exact kind of event will still be causally negative, at the property level, for the kind of effect that actually ensued. Also, of course, the device of focusing on different subpopulations exploited in the previous section is inapplicable in cases like these; that device cannot be used to get token causation and probability increase to agree, for there are no (proper) subpopulations.

So it seems that we cannot expect it to be true of token causation, let alone a characterization of it, that a token cause always rasies the probability of a token effect of it. The question still remains, however, of whether or not there can be *any* probabilistic criterion of token causation. Reason for pessimism is provided by comparison of the squirrel case with the following example, due to Nancy Cartwright: as suggested in Eells and Sober (1983) and Sober (1985), it seems that the two cases are exactly parallel as far as the probabilities go, yet in one case (the squirrel case) we (intuitively) see token causation, and in the other we do not. Consider a normal, healthy plant. The probability of its surviving and being in good health in a year is 0.98. However, Nancy sprayed it with defoliant, and the can said that it is 90 percent effective in killing plants. So the probability of the plant's surviving for a year drops from 0.98 to 0.1. Nevertheless, the plant survives, and it is again healthy in a year. Although the probabilities given in this case are the same as those in the squirrel case, the examples differ in that in the squirrel case we have token causation and in the defoliant case we don't. The squirrel kick token caused the birdie, but the application of the defoliant did not token cause the eventual survival and health of the plant.

The squirrel case shows that it is not crucial to token causation that the kind instantiated by a token cause be a positive causal factor for the kind instantiated by the token effect. Comparison with the defoliant case reinforces the idea that token causation is independent of property level causal relations between the kinds instantiated by the token cause and effect. If our intuitions about the two examples are correct – and I think they are – then it would seem that figuring out what is behind these seemingly conflicting intuitions ("diagnosing" them) would constitute an important step in the direction of understanding (at least our concept of) token causation.

In what follows, I will make a suggestion about what lies behind the asymmetry of our intuitions in the two examples despite the seeming symmetry of the two cases as far as all the probabilities are concerned. This will suggest a characterization of token, or singular, causation. After discussion the strengths and (considerable) limitations of that characterization, I will, in the next section, compare it with the approaches of Rosen, Good and, briefly, that of Salmon (with which my approach seems to share the most in common). In the previous sections, the letters "C" and "E" were used to denote event *types* in a general way. In what follows, it will be necessary to refer to *token* events in a general way. For that, I shall also use the letters "C" and "E", for the context should make it clear whether I am referring to an event type C or E or to a token event of the type C or E.

Unlike causation at the population level, what actually happens after the occurrence of an event C is relevant to whether or not C is a token cause of an event E. For example, whether or not E actually occurs later on is relevant to whether or not C is a token cause of E, for "C token causes E" implies that C and E both actually occur. In addition, I believe that what actually happens *between* C and E in time is relevant to whether or not (at least we would say that) C token causes E. In particular, how *the probability of E evolves* during that time is relevant. And I believe that how we picture $Pr(E)$ (or, in more revealing notation, $Pr_t(E)$) as changing from around the time of C to the time of E marks an important difference between the squirrel and defoliant examples, and explains the asymmetry of our intuitions in the two cases.

The asymmetry of our intuitions derives, I think, from an asymmetry in the way we picture the probability of the effect as evolving in the two examples. In the squirrel case as described (after the kick, the ball is again on a path towards the cup), I think we envision the probability of a birdie as being high just before the kick, then falling abruptly at the time of the kick, but then just as abruptly recovering and becoming high again. See Figure 2. Although the *probability that the probability of a birdie* will take this trajectory is quite low, the probability of a birdie nevertheless does in this case take that trajectory. In the defoliant case, on the other hand, I think we picture the probability of the plant's surviving as slowly creeping back up to near 1 as the plant gradually recovers. See Figure 3. In this case, the application of the defoliant succeeded, for a while at least, in lowering the probability of the plant's surviving. The plant's gradual, but uncertain, recovery is reflected in the slow increase in the probability of survival and eventual health.

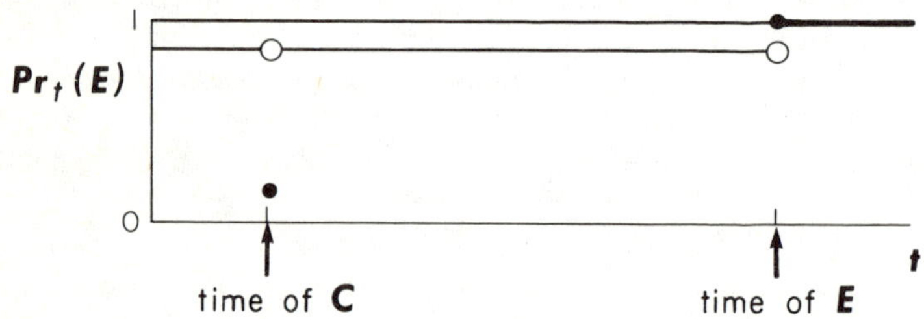

Fig. 2.

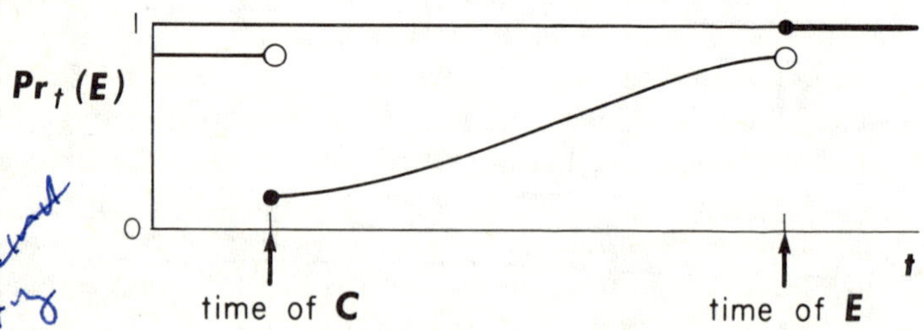

Fig. 3.

For population level causation, a population can always be found within which a causal factor C is either causally positive, negative, or neutral for, a second factor E.[4] That is, C either causes, prevents, or is neutral, for E (in the right population). A parallel classification scheme for the token causal significance of an actual event C for an actual event E (i.e., C token causes, token prevents, or is token neutral for E) is not possible, however. In the first place, any classification of the significance of C for E is complicated by the fact that for C to token cause E, C and E must actually happen, and for C to token prevent E, C and not-E must actually happen. So, in cases like the defoliant example, where C is clearly *relevant* to whether or not E happens (and hence not "neutral" for E), any parallel classification would have to say that C token caused or token prevented E according just to whether or not E happened. But this surely will give us the wrong answer in many cases, for example in our defoliant example.

I believe that for token causation, the appropriate categories of ways in which an event C can be causally significant for an event E are, very roughly and for now intuitively: "E happened *because of* C", "E happened *despite* C", and "E happened *independently of* C", where the last is supposed to indicate that event C played no role in making it probable or improbable that E would happen (perhaps E is outside C's forward light cone). A more comprehensive classification would include a parallel three-way classification for when E fails to obtain, and a final category for the nonoccurrence of C. This is very rough, and a more adequate characterization of the token causal significance of C for E (where both occur) would place C somewhere on a continuum involving because, despite, and independence. Before discussing degrees, I will explain in more detail the extremes of because, despite, and independence, and more carefully distinguish these ideas from the different ways in which one factor can be significant for another at the population level.

An event E happens *independently of* event C if the probability of E does not change around the time of C. An event E happens *because of* event C if (1) the probability of E changed at the time of C, (2) just after the time of C the probability of E is high, (3) this probability is higher than it was just before C, and (4) this probability remains at that high value until the time of E. (Of course it would be possible to specify these conditions in a shorter, more efficient list, but separating them in this way will prove helpful in discussing degrees, below.) And an event E happens *despite* the event C if (1) the probability of E changed at the time of C, (2) just after the time of C the probability of E is low, and (3) this probability is lower than it was just before C (though later on it might go up). These three ways in which an occurrence of C can be causally significant for an occurrence of E are not exhaustive, of course, for they are just the extremes.

The distinction between E's happening because C occurred and E's happening despite the occurrence of C is quite different from the distinction between positive and negative causal significance at the population level, and neither of the former can be characterized in terms of the latter alone (even the latter plus whether or not C and E occurred), nor can the latter be characterized in terms of the former. This is because what happens *after* the occurrence of C – namely, the actual trajectory of the probability of E's happening – is relevant to whether E happens because of or despite C, but is quite irrelevant to whether C is positive, negative, or neutral, for E at the population level. For example, the squirrel case shows that token E can happen because of token C even if type C is a causally negative factor for type E. Thus, negative causal factorhood (together with the actual occurrence of token C and token E) does not imply "despite"; and "because" does not imply positive causal factorhood.

Also, the probability of E's being low just after the occurrence of C (so that token E happens *despite* token C) doesn't imply that C is, at the population level, a negative causal factor for E. Suppose a mediocre golfer strikes the ball and the ball is rolling in the general direction of the cup with a 50 percent chance of falling in. Then a trained monkey kicks the ball exactly in such a way as to confer, at the time of the kick, a 90 percent chance on the ball's falling in. This means that the monkey kick is a positive causal factor for a birdie. However, in about 10 percent of cases exactly like this up until the time of the monkey kick, the ball doesn't fall into the cup, because, let us suppose, the ball in those cases comes off the monkey's foot too far to the right or too far to the left for there to be any significant chance for a birdie. Imagine that in the case in question, the ball does come off much too far to the right for there to be any significant chance of a birdie: the probability of the ball's falling in the cup becomes 0.01, say, and remains low for some time. However, on it's path, the ball is kicked by several squirrels, it ricochets off several trees and boulders, and the wind is just right, all to the effect that a birdie becomes more and more probable as time goes on. Eventually, in this case, the ball, improbably enough, lands in the cup. In this case, the birdie occurred *despite* the actual token monkey kick (an event whose *actual* "effect" was a *decrease* in the probability of a birdie), even though events of just that kind are, at the population level, causally positive for a birdie.

The because/despite distinction has to do with the trajectory of the probability of the later event E *after* the occurrence of the earlier event C, and not with the probability of E *at the time of* C. I think the population level theory of probabilistic causality correctly identifies the probability of E *at* the moment of C, but that the probability of E at this moment is irrelevant to whether or not C token causes E in particular cases (except to the extent that raising or lowering the probability of E at one time is relevant to its being high or low at later times, on which see below). What is relevant to probabilistic token causation, it seems to me, is the trajectory of the probability of E after the moment of C. That this is so, and the plausibility of my characterizations of the extremes of because and despite in terms of the trajectory of the probability of E after C occurs, is strongly supported, I think, by the plausibility of my diagnosis of our conflicting intuitions in the squirrel and defoliant cases.

There are various ways in which an occurrence of an event C can be significant for an event E without being at an extreme of because, despite, or independence. E can happen "more or less" because of C, and "less or more" despite or "less or more" independently of C, depending on the "degree" to which the various clauses in the definition of "because" are satisfied (and there can be tradeoffs other than those indicated by the order of the words "less" and "more" above). To see this more easily, let's suppose for a moment that the probability of E doesn't change between just after C and the time of E. (See Figure 4 for a diagrammatic depiction of some of the following possibilities.) Assuming that the probability of E is high just after C, and that it's higher after than before C, then where the significance of C for E falls, between "because" and "independence", depends on the magnitude of the difference between the probability of E just before C and the probability of E just after C. And assuming that the probability of E is low just after C, and that there is a decrease in this probability from what it is before to what it is after C, the significance will fall between "despite" and "independence" at a point determined again by the absolute magnitude of the change of the probability of E across the time of C.

These two kinds of nonextreme causal significance seem, at least to

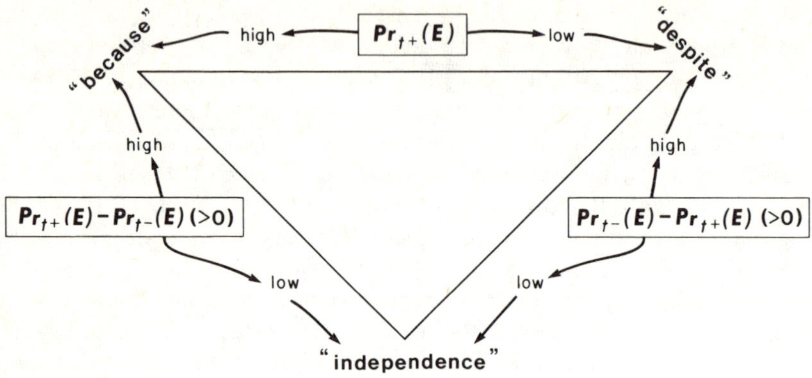

Fig. 4.

me, quite intuitive. But also, assuming any given change in the probability of E, in either direction, it seems that the significance of C for E should lie closer to because than to despite to the degree to which the probability of E is high just after the time of C. For example, suppose that after the squirrel kick the probability of a birdie was not quite 0.98 (where it was before the kick), but only 0.95; I think our intuitions remain committed to the idea that it was the squirrel's kick that caused the birdie.

Besides all this, there is another reason why the causal significance of an occurrence of C for one of E should be classified more or less as of the "independence" variety. Return to the original simple squirrel example. Now suppose that just after the squirrel's kick, the probability of a birdie is 0.99 (so that probability has moved from 0.98 just before the kick to 0.1 at the time of the kick, to 0.99 just after the kick). If this probability remains unchanged until the time of the birdie, then we have causal significance of the "because" variety. However, suppose now that some time after this squirrel's kick, a second squirrel kicks the ball and the probability of a birdie decreases to 0.1, until, at a later time, a third squirrel kicks the ball and the probability of a birdie then increases to 0.99, and then stays there up until the time of the actual birdie. In this case, I think we should say that the birdie happened more or less independently of the first squirrel kick.

In general, therefore, it seems that where the causal significance of an occurrence of an event C for the occurrence of an event E falls on the three dimensional because/despite/independence continuum depends on: (1) the sign of the difference between the probability of E before and after the occurrence of C, (2) the magnitude of that difference, (3) beyond that, the value of the probability of E after the occurrence of C, and, as the last example illustrates, (4) just how the probability of E changes after C – e.g., how much, in what directions, and how many times.

The characterization of token causation just given is not without its weaknesses and limitations, of course. In the first place, of course, it is formally quite incomplete and imprecise – just the beginning of a "real" theory. For another thing, the trajectory of the probability of E between the time of C and the time of E does not completely reliably indicate the forces that may be impinging upon that trajectory. For example, if according to the above characterization, E happens because of C, but there are downward forces on the trajectory of the probability of E that are exactly compensated by simultaneous upward forces, then, to the extent to which there are such forces, it would seem that C loses some of its positive (or "because" variety) token causal significance for E. Another example of this kind was pointed out to me by Igal Kvart: suppose that at the same time that a lit match is thrown into a dry barn, a bucket of water is emptied in the field. The theory above can't tell the token causal significances of these two events for the fire apart. Thus, it seems clear that considerations beyond probabilistic features of a situation will be essential to any adequate and complete analysis of token causation.

There also, of course, is the question of what the appropriate interpretation of probability is for the theory. In particular, there is the question of exactly what the probability trajectories used in the theory are supposed to represent. For the squirrel case, for example, it was supposed that the probability of a birdie dropped sharply *at the time of the kick*. Thus, when plotting the trajectory of the probability of a birdie, we should ignore factors such as the ball's getting closer and closer to the squirrel, making it more and more probable that it will kick the ball and prevent a birdie. And in the defoliant example, we should ignore factors such as the can of defoliant getting closer and closer to the plant, making it already less and less probable that the plant will be alive and healthy in a year. If we don't ignore such factors, then the above characterization would say that the plant's survival was *independent* of the application of the defoliant. Roughly, what the trajectory represents is the probability, at times, of the later event (E) on the assumption that all factors relevant to whether or not the effect will happen have already fallen into place. Given a clarification of this idea, perhaps then the "laws of motion" of probabilities may be more adequately investigated and clarified.

In addition, I think the above discussion only gives a formalization of *our intuitions*, controlling what we would say in certain situations. It is a theory about how we label the causal significance of a C for an E, as a function of how we picture the trajectory of the probability of E from around the time of C to the time of E. It is about *our concept* of token causation. That is, it deals more with the *epistemology* of token causation than with the *metaphysics* of token causation. The latter would deal with important questions left untouched here. For example, what is the significance of the seeming conflict between the fact that 1) the nature of changes in the trajectory of the probability of E after C and until E (and whether or not there are any such changes) is completely out of the control of C and 2) the trajectory of this probability in that interval is (at least according to my characterization above) relevant to whether or not C token causes E in the situation in question. Another puzzling feature of the analysis is that it is far from clear that where the probability of E winds up just after the occurrence of C can always be said to be exactly a *causal effect* of C (especially if it winds up at a value quite different from that of the probability of E *at* the time of C); yet where it winds up then *is* relevant (as a kind of intermediate, yet perhaps not quite *causally* intermediate feature of the situation) to whether or not C token causes E.

Support for my characterization of singular causation is provided by the plausible diagnoses, provided by the theory, of various puzzling intuitions we have in certain examples. The epistemological resolution of those puzzles provided by the theory results, however, in puzzles about the relationship between the epistemology and the metaphysics of singular causation. It is far from crystal clear to me whether these puzzles provide a "reductio" of my characterization of the epistemology of singular causation, a "reductio" of the originally puzzling intuitions which the theory attempts to accommodate, or perhaps some insight into the nature of singular causation itself. In any case, it is clear, as pointed out above, that the analysis given is only a skeletal outline that might turn out to be a more adequate theory when supplemented with whatever relevant ideas beyond mere probabilities may turn out to be necessary.

4. COMPARISON WITH OTHER APPROACHES

In this section, I will examine three other main kinds of approach to the problems raised in the previous section for characterizing token causation probabilistically, and compare these approaches with the one outlined in the previous section.

Deborah Rosen has suggested that the theory that causes raise the probability of their effects can be saved if we relativize the definition of probabilistic causation to "background contexts". Recall Rosen's golfing example. The golfer swings, the ball hits a limb of a tree, and then it falls into the cup. The ball's hitting the tree caused the birdie, but a birdie is less probable given that the ball hits a tree than given that it doesn't. In this case, the background context would include information about "the angle and the force of the approach shot together with the deflection [by the tree limb]" (1978, p. 608). Including this information results in a "broader", "revised" causal picture of the situation, and "the results [the birdie in this case] are unlikely only from a narrow standpoint" (p. 608).

Rosen's suggestion, then, is, roughly: if we characterize a token cause in enough detail, then events of just the same (actually "narrower") kind as the token cause will increase the probability of an effect of the kind that actually occurred. Salmon (1980, 1984) has expressed scepticism about this approach, which he calls "the method of more precise specification of events", pointing out that "whether the ball will drop into the hole is extremely sensitive to minute changes in the conditions of the collision" and that "an unrealistically detailed description of the surface texture of the branch would be required to yield even a reasonable probability for the hole-in-one [*sic*]" (1984, pp. 194–195). I am inclined to agree with Paul Humphreys (1980) that this reason for rejecting Rosen's approach is "more appropriate to the pragmatics of explanation than [to] finding the causes of the birdie" (p. 311), but I also agree with both Salmon and Humphreys that the approach is nevertheless not generally applicable. Just consider the squirrel example above, in which the squirrel kick confers a physically *irreducibly* low statistical probability on the ball's falling into the cup – and lower than it would have been without the kick.

It has been pointed out by a number of authors (e.g., Suppes 1970; Rosen 1978; Salmon 1980, 1984; and Otte 1981) that, in Rosen's example, before the ball hits the limb, it is moving in such a way that, were it not to hit the limb, it would almost certainly overshoot the cup.

Thus, given the way the ball is moving before striking the limb, its hitting the limb provides just about the only chance of its landing anywhere near the cup, and just about the only chance of its falling in. Thus, taking into account the fact that the ball would almost certainly miss the cup if it didn't hit the tree limb, we see that its hitting the tree limb actually does raise the probability of a birdie, where of course what is relevant is that there is a probability *increase* and not that there be an increase *to close to 1*. Again, however, this is hardly a general feature of situations in which a cause seems to lower the probability of its effect. Consider again the squirrel example, in which the ball was moving straight towards the cup, with a high probability of falling in, before the squirrel came into the picture.

Notice that in Rosen's approach, the sense in which the probability of an effect E is supposed to be increased by the occurrence of a cause C is not simply that the probability of E in fact *winds up* higher *after* C: her approach is not about the actual trajectory of the probability of the effect, but rather about the probability that *C confers on* E, that is, the probability of *E at the time of* C. The discussion in the previous section of this paper makes it clear that we can't expect these two values always to coincide. I suppose that in the most likely (at least most easily pictured) case of Rosen's example, the probability of a birdie just *after* the ball's collision with the tree limb – when the new course of the ball has become established – is rather high, even if the effect of the collision (the new course of the ball) is indeterministic in nature, so that, at the time of collision, the probability of a birdie may still be quite low. In Rosen's example, this latter value may be higher than the probability of a birdie given that the ball misses the tree (but is heading in the direction of the tree and thus in a direction away from the cup), but this is a feature peculiar to Rosen's particular example. By looking at the actual trajectory of the probability of the effect (and ignoring the question of what the cause's role might have been in producing that particular trajectory), the theory of the previous section of this paper comes closer, I think, to preserving, in a plausible and general way, the idea of probability increase in token causation.

I. J. Good (1961–2) was, as far as I know, the first to recognize the distinction between, as he puts it, "the tendency of C to cause E" and "the degree to which C actually caused E".[5] And he gives analyses of both ideas. For the tendency of C to cause E, Good proposes, basically, a measure of statistical relevance, and his analysis of this first concept can be viewed as a quantitative version of the standard theory of probabilistic causality sketched at the beginning of this paper, except that only one background context is considered, the one that actually obtains (including the true laws of nature). Thus, Good's analysis of the tendency of C to cause E supplements a version of the qualitative population level theory of causation with a quantitative measure of *degree* of positive or negative causal significance. Here, however, I shall be mainly concerned with Good's analysis of the degree to which C actually caused E.

Recall Good's example involving Holmes, Moriarty, Watson, the cliff, and the boulder. In this example, Watson's pushing the boulder off the cliff had a *negative tendency to cause* Holmes' death by crushing, but nevertheless *actually caused* Holmes' death by crushing. Good says, "F [Watson's push] had a tendency to prevent E [Holmes' death by crushing] and yet caused it. We say that F was a cause of E because there was a chain of events connecting F to E, each of which was strongly caused by the preceding one" (1983, pp. 216–217). The degree to which C actually causes E is a measure of strength of token causation. And this degree is high to the extent to which links of a chain of actually occurring events from C to E are, basically, highly positively probabilistically relevant to their successors (or, better: events of each of the successive kinds that were actually instantiated have a positive *tendency* to cause – are highly causally positive for at the properly level – events of the kinds of their successors).

The essence of Good's theory can, I think, be expressed as follows. C token causes (or caused) E if there is a chain of events – $C, F_1, \ldots, F_n, E$ – such that (1) C, E, and the F_i's actually occurred; (2) they are spatio-temporally "adjacent" without "overlapping" very much; and (3) each has a strong *tendency* to cause its successor (at the population or property level, each is strongly causally positive for its successor). And the *degree* to which C actually causes E (a number ≥ 0) is a function of the strengths of the individual links in the chain.[6]

Good's theory and the theory sketched in the previous section give different answers to the question of whether or not one event token causes another in certain situations. Consider again the case of the sprayed plant. Recall that the application of the defoliant reduces the chances of survival to 0.1, but the plant gradually recovers and survives, and the probability of survival gradually recovers in time. The theory of the previous section of this paper says that the plant survived *despite* the application of the defoliant (since the probability of survival became quite low after the spraying and stayed low for a while). And this seems to me to be the correct evaluation: the application of the defoliant did not *cause* survival, the plant did not survive *because* of the application of the defoliant. However, it seems likely that there exists a chain of events connecting the application of the defoliant to the plant's later good health, such that each event in the chain is strongly causally positive for its successor. The spraying has a strong tendency to cause the presence of the defoliant on the plant's surface; this has a strong tendency to cause the plant to be unhealthy in some particular way; this has a strong tendency to cause the plant to react in a certain way, setting into motion a process of combatting the deleterious effects of the poison; etc. (Whether or not this is biologically quite accurate is not, of course, to the point; surely there are *bona fide* examples that illustrate the point, perhaps involving a body's immune system's reaction to a typically deadly disease.) So Good's analysis seems to imply that the spraying actually caused the plant's survival, indeed to a degree that is high to the extent that the connections between successive members of the chain are strong.

There are two key features of the difference between Good's theory and that of the previous section of this paper that allow the latter to give what I think is the right answer but make Good's theory give what I think is the wrong answer in cases such as this. First, the theory of the previous section focuses at all times on the probability of E, rather than on the relevance of one event in a chain from C to E to the next event in the chain. It is possible, of course, to put a lot of strong connections together in a chain and still wind up with a rather weak connection between the first and last elements of the chain. The model of the previous section reflects this feature of the defoliant case by the *gradual* increase in the probability of survival: as time goes on and more and more of the strong connections in the chain are successfully made, the higher the probability of survival becomes, because there are fewer of the strong connections left that have to be made for the plant to survive. Although Good's theory is sensitive to the fact that a series of strong connections may itself be a weak connection, it is important to note that the strength of a chain from an event C to an event E is, on Good's analysis, independent of the actual initial probability of E (i.e., at the time of C) so that, for example, this strength (the degree to which the C actually caused E) can be the same whether the trajectory of the probability of E is high all the way from the time of C to the time of E, or gradually recovers from a low value at the time of C to a large value at the time of E. (This feature of Good's analysis is illustrated, incidentally, by Salmon's example involving Joe Doakes and Jane Bloggs, 1980, pp. 54–55.)

Another key difference between Good's theory and that of this paper is that the latter includes the category of "despite", while the former does not. In Good's theory, the strength of a causal connection between an event C and a later event E is a number $\geqslant 0$. So, intuitively, and in terms of the terminology of last section, Good's theory allows for the extreme of E's happening *independently of* C and for degrees to which E happens *because of* C. To capture the token causal truth in cases like the defoliant example, however, it seems that a third dimension is needed. Surely the plant didn't survive *because* of the spraying; and surely the spraying had *some* causal role with respect to the survival of the plant, that is, the survival of the plant was surely not *independent of* the spraying. That is why the theory of the previous section of this paper avails itself of a third category, that of "despite". Given the way "despite" is characterized in the previous section, we can say that one way in which the first difference between the two theories noted above shows up is in this second difference.[7]

The two theories also give different answers in the case of the squirrel kick. Just after the kick, recall, the ball's momentum and the positions of obstacles on the course, etc., were, *improbably enough*, such that the probability of a birdie became very high again. But that was *improbable*: there was a very strong tendency of the squirrel kick to *prevent* this. Thus, Good's theory says that the degree to which the squirrel kick actually caused the birdie was very low. (On Good's theory, a chain is no stronger than its weakest link.) The theory of the previous section, however, categorizes the significance of the kick for the birdie more or less at the "because" extreme. The difference between the two theories that makes the difference in cases of this kind is that while Good's uses just the type level probabilistic relations in terms of which property level causal tendencies are characterized, the theory of the previous section uses the actual trajectory of the probability of the effect E, regardless of what the causal role or tendency might have been of the cause E, or of intermediate events, to *produce* just the trajectory that the probability of E actually followed.

Wesley Salmon has called the approach of invoking a causal chain and examining the individual connections in it "the method of successive reconditionalization" (1980, 1984). Salmon has objected that there *may not be* any intermediate factors between C and E that are strongly causally connected to their immediate successors and predecessors. He considers a fictitious case of an atom that can be in one of four states, called "1", "2", "3", and "4". It started in state 4. The probability of its going into state 3 is 0.75, and into state 2 is 0.25. If it goes into state 3, then its probability of going into state 1 next is 0.75; but if it goes into state 2 from 4, then its probability of going into state 1 next is 0.25. (See Figure 5.) What actually happens, however, is that the atom goes, improbably enough, from state 4 to 2 to 1. Being in state 2 is negatively relevant to going into state 1, yet according to Salmon, this is what actually caused it to do so. And note that there is no way to fill in intermediate links so as to "mend" the simple 4–2–1 chain.

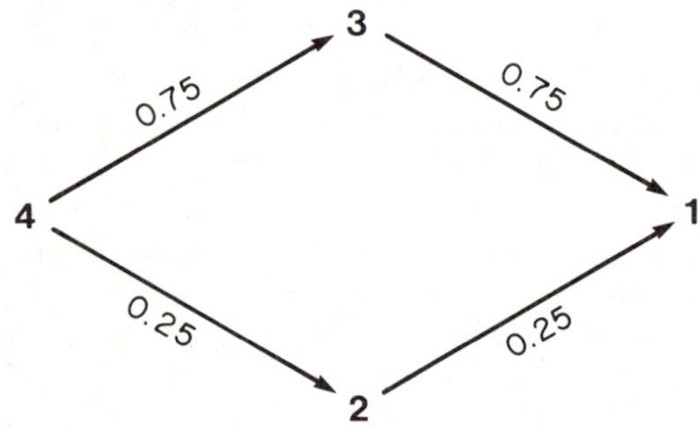

Fig. 5.

I agree, of course, that there are cases in which C token causes E but in which C is negatively probabilistically relevant to E and in which there are no intermediate events that can be interpolated to "mend the chain": the squirrel case, for example. However, I don't think that Salmon's example is most naturally construed as such a case. Although there would seem to be no interesting "trajectory" of the probability of going into state 1 between the time the atom is in state 2 and the time it is in state 1, it seems to me more natural to say that the atom went into state 1 *despite* its having been in state 2.[8]

In any case, Salmon adopts the approach of dropping the idea of *classifying* token causal significances as (more or less) positive, independent, or whatever, favoring the idea of taking as basic simply the "transmission of a probability distribution" through time, i.e., the idea of what he calls "causal processes", which can be modified by "interactions", or "intersections" with other causal processes. An event is an intersection of one process with another, and on Salmon's approach, the token causal significance of one event for a later event can only be given by a description of the evolution of probability distributions carried by (or constituting) the relevant processes.

The idea of a causal process is close, I think, to the idea used in this paper of the trajectory of the probability of an event. But I have argued that, when the categories of "despite" and "independence" are allowed (to accommodate the chancy nature of the connection between one event the ensuing trajectory of the probability of another), categorization of token causal significances may plausibly be reintroduced.

University of Wisconsin-Madison

NOTES

* I presented this paper to the National Science Foundation conference on Probability and Causation, Irvine, California, July 15–19, 1985. I thank Elliott Sober for helpful discussions before the meeting, and a number of those in attendance at the meeting – especially William Harper, Igal Kvart, Elizabeth Lloyd, Fred McGifford, and Wesley Salmon – for useful criticism and suggestions. And I thank the American Council of Learned Societies for financial support.

[1] See Good (1961–2), Suppes (1970), Cartwright (1979), Skyrms (1980), and Eells and Sober (1983) for versions of the theory, and especially the last for a number of qualifications. The version stated in the text is virtually identical to Cartwright's formulation. Note the circularity of the theory; Cartwright believes the circularity is essential, that the kind of causation here characterized cannot be defined in terms of regularities, or probabilities, alone.

[2] See Dupré (1984) and Eells (1987a) for further discussion.

[3] Otte himself suggests a treatment like this in his (1981, p. 182–183). For further discussion, see my (1987b).

[4] The reason for the qualification about finding the "right" population is that "causally positive", "causally negative", and "causally neutral" are defined in terms of the ideas of "always raises", "always lowers", and "always leaves unchanged" the probability of the effect, respectively. While mutually exclusive, these three ways in which one factor may be significant for another are not exhaustive. See above, section 2, and Eells (1986) for further discussion, and the appropriate "methodological" approach in situations in which none of the three kinds of significance applies.

[5] Good uses "*F*" instead of "*C*". But for uniformity of notation, I'll continue to use "*C*" for the cause.

[6] Actually, Good's chains are a special case of his "nets". And the degree to which C actually causes E is actually defined as the *limit* of the strength of nets connecting C to E when the nets gets more and more fine grained. But these details do not affect the points I make below.

[7] Wesley Salmon suggests that "despite" may be characterizeable in terms of Good's idea of "resistance", the complement of the "strength" of a chain.

[8] Good makes a point similar to this in his (1980), and Salmon seems to sympathize with the idea (1980, note 24; but also 1984, p. 202).

REFERENCES

Cartwright, N. (1979) 'Causal laws and effective strategies', *Noûs* **13**, 419–437.

Dupré, J. (1984) 'Probabilistic causality emancipated', in *Midwest Studies in Philosophy IX: Causation and Causal Theories*, P. A. French, T. E. Uehling, Jr., and H. K. Wettstein (eds.). (Minneapolis: University of Minnesota Press.) Pp. 169–175.

Eells, E. (1986) 'Probabilistic causal interaction', *Philosophy of Science* **53**, 52–64.

Eells, E. (1987a) 'Probabilistic causality: Reply to John Dupré', *Philosophy of Science* **54**, 105–114.

Eells, E. (1987b) 'Cartwright and Otte on Simpson's Paradox', *Philosophy of Science* **54**, 233–243.

Eells, E. and Sober, E. (1983) 'Probabilistic causality and the question of transitivity', *Philosophy of Science* **50**, 35–57.

Good, I. J. (1961–2) 'A causal calculus I–II', *British Journal for the Philosophy of Science* **11**, 305–318; **12**, 43–51; *Errata and Corrigenda*, **13**, 88. Reprinted in Good (1983), pp. 197–217.

Good, I. J. (1980) 'Some comments on probabilistic causality', *Pacific Philosophical Quarterly* **61**, 301–304.

Good, I. J. (1983) *Good Thinking* (Minneapolis: University of Minnesota Press).

Good, I. J. (1985) 'Causal propensity: A review', in *PSA 1984*, Vol. 2, P. D. Asquith and P. Kitcher (eds.). (East Lansing, Michigan: Philosophy of Science Association.) Pp. 829–850.

Hesslow, G. (1976) 'Discussion: Two notes on the probabilistic approach to causality', *Philosophy of Science* **43**, 290–292.

Humphreys, P. (1980) 'Cutting the causal chain', *Pacific Philosophical Quarterly* **61**, 305–314.

Otte, R. (1981) 'A critique of Suppes' theory of probabilistic causality', *Synthese* **48**, 167–189.

Otte, R. (1985) 'Probabilistic causality and Simpson's paradox', *Philosophy of Science* **52**, 110–125.

Rosen, D. A. (1978) 'In defense of a probabilistic theory of causality', *Philosophy of Science* **45**, 604–613.

Salmon, W. C. (1980) 'Probabilistic causality', *Pacific Philosophical Quarterly* **61**, 50–74.

Salmon, W. C. (1984) *Scientific Explanation and the Causal Structure of the World*. (Princeton, New Jersey: Princeton University Press.)

Skyrms, B. (1980) *Causal Necessity*. (New Haven: Yale University Press.)

Sober, E. (1985) 'Two concepts of cause'. In *PSA 1984*, Vol. 2, P. D. Asquith and P. Kitcher (eds.). East Lansing, Michigan: Philosophy of Science Association. Pp. 405–424.

Suppes, P. (1970) *A Probabilistic Theory of Causality* (Amsterdam: North Holland Publishing Company).

PATRICK SUPPES

PROBABILISTIC CAUSALITY IN SPACE AND TIME

It is my intention in this paper to survey results and problems in the theory of probabilistic causality with an emphasis on the theory in the context of space and time. In view of this emphasis, the first section is devoted to problems of space and the second to problems of time. Here, space and time are construed in the sense of classical physics, although in some of the examples considered no real physics will enter. The third section is devoted to space-time, but the problems considered are restricted to those that arise in the framework of special relativity. Questions about probability and causality are difficult enough in this framework without considering the still more difficult case of general relativity.

As in inevitable in such surveys, the work has too much emphasis on problems that I have been concerned with myself. I therefore lay no claim to offering a balanced objective overview, but only one of some topics that I think are of general interest.

1. SPACE

The fundamental problem of probabilistic causality in the framework of classical space can be formulated this way: *explain simultaneous but distant correlated events*. The general methodology goes back at least to Reichenbach and can be summarized under the slogan: find a common cause.

I begin with a theorem that shows how intimately the existence of a common cause is related to the joint distribution of random variables. There are two things to note about the theorem. First, it is restricted to random variables that are two-valued – the kind of random variable that would ordinarily be used to represent a qualitative property rather than a quantitative one. Second, the criterion for the existence of a common cause that is generally accepted is the conditional independence of the random variables given the common cause. Note that in the case of two-valued random variables this can be expressed just in terms of the conditional expectation of the random variables. For

Brian Skyrms and William L. Harper (eds.), Causation, Chance, and Credence, Vol. I, 135–151.

random variables with more values, considering just their expectations will not be sufficient to express their conditional independence.

THEOREM 1 (Suppes and Zanotti 1981). *Given phenomenological random variables* $X_1, \ldots, X_N$, *then there exists a common cause* λ *such that*

$$E(X_1, \ldots, X_N \mid \lambda) = E(X_1 \mid \lambda) \cdots E(X_N \mid \lambda)$$

if and only if there exists a joint probability distribution of $X_1, \ldots, X_N$. *Moreover,* λ *can be constructed as a deterministic cause, i.e., for* $1 \leqslant i \leqslant N$, *the variance of* X_i *given* λ *is zero.*

What is important about this theorem is that the only condition on the existence of a common cause is the phenomenological condition of having a joint probability distribution.

We also then have the following corollary.

COROLLARY. *Any joint distribution of* $X_1, \ldots, X_N$ *can be represented as a mixture of conditionally independent distributions:*

$$F(X_1, \ldots, X_N) = \int_{-\infty}^{\infty} F_1(x_1 \mid \lambda) \cdots F_N(x_N \mid \lambda)\, \mathrm{d}G(\lambda).$$

There are various consequences of the theorem that are worth examining but that I will not formulate in formal corollaries. Some applications are examined.

Quantum mechanics. By virtue of Theorem 1 the existence of hidden variables in the sense of classical quantum mechanics is equivalent to the existence of a joint probability distribution for two-valued random variables. This fact has not been sufficiently emphasized in foundational discussions of hidden variables.

Bell (1964) derived his well-known equations as phenomenological conditions on the existence of a hidden variable. The equations were put in better form by Clauser *et al.* (1969). Most surprising of all, Fine (1982) showed that Bell's inequalities were a sufficient condition for the existence of a joint probability distribution. It is easy to show that Bell's inequalities are a necessary consequence of the existence of a joint probability distribution.

To be technically more explicit, these matters may be summarized in the following way. Bell's inequalities are specifically formulated for measurements of spin of pairs of particles originally in the singlet state. Let A and A' be two possible orientations of apparatus I, and let B and B' be two possible orientations of apparatus II. Let the measurement of spin by either apparatus be 1 or -1, corresponding to spin 1/2 or $-1/2$, respectively. By $E(AB)$, for example, I mean the expectation of the product of the two measurements of spin, with apparatus I having orientation A and II having orientation B. By axial symmetry, we have $E(A) = E(A') = E(B) = E(B') = 0$, i.e., the expected spin for either apparatus is 0. It is, on the other hand, a well-known result of quantum mechanics that the covariance term $E(AB)$ is:

$$E(AB) = -\cos \theta_{AB},$$

where θ is the difference in angles of orientation A and B. Again, by axial symmetry only the difference in the two orientations matters, not the actual values A and B. (To follow the literature, we begin with the notation A, B, A' and B' for phenomenological random variables, rather than $X_1, \ldots, X_N$, which we go back to later.)

On the assumption that there is a hidden variable that renders the spin results conditionally independent, i.e., that there is a causal hidden variable λ in the sense of Theorem 1, Bell (1964) derives inequalities equivalent to the following.

$$(1) \quad \begin{cases} -2 \leqslant -E(AB) + E(AB') + E(A'B) + E(A'B') \leqslant 2, \\ -2 \leqslant E(AB) - E(AB') + E(A'B) + E(A'B') \leqslant 2, \\ -2 \leqslant E(AB) + E(AB') - E(A'B) + E(A'B') \leqslant 2, \\ -2 \leqslant E(AB) + E(AB') + E(A'B) - E(A'B') \leqslant 2. \end{cases}$$

This form of the inequalities is due to Clauser *et al.* (1969).

THEOREM 2 (Fine 1982). *Bell's inequalities are necessary and sufficient for the existence of a joint probability distribution compatible with the six given covariances of the phenomenological random variables A, A', B, and B' occurring in the inequalities* (1).

On the other hand, pursuing these results about hidden variables, it is natural to ask what is the causal situation when more random variables are considered. We know from much experience with conditions on joint distributions that the conditions must expand with the number of variables, i.e., it would be the most improbable event imaginable to have simply a fixed condition on subsets of four random variable be sufficient for a joint distribution of all the random variables. We know that we must consider subsets of increasingly large size and nonreducible conditions on these larger and larger subsets. This general experience is well borne out here. Garg and Mermin (1982) have given a counterexample to Bell's inequalities being sufficient for eight random variables when what we term *Bell covariances* are given. By Bell covariances we mean covariances $E(X_iX_j)$ for $1 \leqslant i \leqslant i_0 < j \leqslant N$, for some integer i_0. Garg and Mermin's counterexample is for $N = 8$ and $i_0 = 4$. Let $E(X_1X_5) = E(X_2X_6) = E(X_3X_7) = 1$ and otherwise $E(X_iX_j) = -1/3$ for $1 \leqslant i < j \leqslant N$. Then it is easy to show that for the quintuple $(X_1, X_3, X_4, X_6, X_8)$ all covariances must be $-1/3$. But it follows at once from Theorem 3 stated below that with $N = 5$, $a_i = a_j = 1$, for existence of a compatible joint distribution $\Sigma E(X_iX_j) \geqslant -2$, and so there can be no joint distribution compatible with the given covariances all equal to $-1/3$.

It is natural to ask more generally under what conditions given covariances of pairs of phenomenological random variables are such that they are compatible with the existence of a joint distribution of the random variables. Fortunately, a rather simple condition can be given, although the proof of its correctness is rather intricate.

THEOREM 3 (Suppes and Zanotti, to appear). *A necessary and sufficient condition that there exist a joint probability distribution compatible with the given covariances of all pairs of N phenomenological random variables is that*

$$\sum_{i<j} a_i a_j E(X_iX_j) \geqslant (1 - n)/2$$

for all subsets of odd cardinality $n \leqslant N$ *and with* $a_i, a_j = \pm 1$.

The various hidden-variable theorems relevant to quantum mechanics have been formulated with the weakest possible conditions on the causal variables. The reason is certainly because everyone has expected that quantum mechanics and relevant experimental data would probably support quantum mechanics. Therefore, the strongest possible test of quantum mechanics versus causal theories with hidden variables is to impose the weakest possible conditions on the causal theories, and still show their lack of empirical support.

On the other hand, it is also of interest to try to ask what are reasonable conditions on causal hidden variables in order to make them physically acceptable. Without restrictions, highly artificial mathematical constructions can be given, as is the case for the deterministic hidden variables of Theorem 1.

An example of such a theorem is one that imposes some natural symmetry conditions both at a phenomenological and at a theoretical level. The main principle of symmetry I shall use is that of exchangeability. Two random variables X and Y with possible values ± 1 are said to be *exchangeable* if the following probabilistic equality is satisfied:

$$P(X = 1, Y = -1) = P(X = -1, Y = 1).$$

The theorem I state shows that if two random variables are exchangeable at the phenomenological level then there exists a hidden causal variable satisfying the additional restriction that they have the same conditional expectation if and only if that correlation is nonnegative.

THEOREM 4 (Suppes and Zanotti 1980). *If X and Y are exchangeable, then three exists a hidden variable λ such that*

(i) $$E(XY|\lambda) = E(X|\lambda)E(Y|\lambda)$$

(ii) $$E(X|\lambda) = E(Y|\lambda)$$

if and only if

$$\rho(X, Y) \geqslant 0.$$

Certainly it is possible to criticize this theorem as being too strong in its symmetry requirements from a physical standpoint. This has been done by Shimony (1981). More can be said about these matters but in this survey it seems more useful to turn to entirely different fields of investigation where similar ideas still apply.

Social sciences. Among the most widely used models in the social sciences are latent trait models. The applications range from the theory and analysis of mental tests to the analysis of sociological surveys on a variety of attitudes toward topics of political or social importance. Let us take ability in the case of mental tests and attitude in the case of surveys. In such applications, the latent trait represents an ability or attitude measured by the test or survey instrument. The data consist of individuals' responses to the test or survey. It is understood that the responses are of a yes–no variety for the purposes of the present discussion. A latent trait model is in principle just like a hidden-variable model in quantum mechanics. The objective is to identify theoretical variables that render the probabilities of responses to the various items of the test or survey conditionally independent, i.e., conditionally independent given the theoretical trait variable. It is the job of the latent trait model to predict the observed probabilities, and to explain the phenomenological correlation between items by the standard common cause approach, i.e., by rendering the items conditionally independent.

Theorem 1 shows that if no restrictions are placed on latent trait models then as long there is a joint probability distribution such latent traits can always be found. This fact has not at all been generally recognized in the extensive literature on latent trait models. The theory has not been oriented toward general causal questions as in the case of quantum mechanics, but to the testing of specific models. Unlike the case of quantum mechanics, in most ordinary applications of latent trait models one expects the test or survey, simply on the principles of experimental design, to have a proper joint distribution of item responses. So in principle no problem arises of finding hidden variables, and the very weak theory, interesting because of the expected negative results in the case of quantum mechanics, does not admit of the same interest in this social science context. Holland and his associates (Holland 1981, Cressie and Holland 1983, and Rosenbaum 1984) have studied when various latent trait models are consistent with observed data. In this case, additional restrictions are placed on latent traits. For example, one standard restriction is that the conditional distribution of an item must always be monotone in the latent trait variable λ.

Some technical details are needed to formulate the ideas explicitly. I follow the excellent general analysis of these matters to be found in Holland and Rosenbaum (1986). First, we say that the distribution of a random vector $\mathbf{X}$ is *associated* if $\mathrm{Cov}(f(\mathbf{X}), g(\mathbf{X})) \geqslant 0$ for all non-decreasing, bounded functions f and g of $\mathbf{X}$. Obviously this is a concept of positive association for the components of $\mathbf{X}$. (For notational purposes it is simpler to refer to the random vector $\mathbf{X}$ rather than the unidimensional variables $X_1, \ldots, X_N$.) Second, the distribution of a random vector $\mathbf{X}$ *is conditionally associated* if, for any partition $(\mathbf{Y}, \mathbf{Z})$ of $\mathbf{X}$ and any function $h(\mathbf{Z})$, the conditional distribution of $\mathbf{Y}$ given $h(\mathbf{Z})$ is associated. Third, *latent conditional independence* is just the kind of independence expressed in Theorem 1 but generalized to more than two values. In particular, it is the condition that

$$F(x_1, \ldots, x_N | \lambda) = \prod_{j=1}^{N} F_j(x_j | \lambda)$$

for all $x_1, \ldots, x_N$ and λ, where F is the conditional distribution of $x_1, \ldots, x_N$ given λ. Fourth, a latent variable model $(\mathbf{X}, \boldsymbol{\lambda})$ satisfies *latent motonicity* if the functions $1 - F_j(x|\lambda)$, $j = 1, \ldots, N$ are all nondecreasing functions of λ for all x. In case λ is a vector, $1 - F_j(x|\lambda)$ must be nondecreasing in each of the coordinates of λ. Fifth, a model $(\mathbf{X}, \boldsymbol{\lambda})$ is *unidimensional* if $\boldsymbol{\lambda}$ is a scalar, i.e., is unidimensional.

THEOREM 5 (Holland and Rosenbaum 1986). *If a latent variable model* $(\mathbf{X}, \lambda)$ *satisfies the conditions of latent conditional independence, latent monotonicity and unidimensionality then the distribution of X is conditionally associated.*

In contrast to Theorem 1 the three conditions on the common cause $\boldsymbol{\lambda}$ given in Theorem 5 impose strong phenomenological conditions of positive association on the observable variables.

It is not always the case that joint distributions of the phenomenological variables are guaranteed by experimental design in test theory. For a discussion of the problems that are due to practice effects, see Holland and Thayer (1985). Let $\mathbf{X}$ be a random vector representing a test score and $\mathbf{X}_p$ the test score after taking a practice test. Holland and Thayer point out that we cannot collect data in any direct way on the joint distribution of $\mathbf{X}$ and $\mathbf{X}_p$.

There are a variety of other examples in which various general restrictions are imposed. These general restrictions usually make a good deal of intuitive sense and thus render the search for a common cause empirically testable and interesting.

Distance as a cause. One kind of standard model of spread of epidemics makes the fundamental assumption that the probability that an individual becomes infected is monotonically decreasing in distance from the source of infection. This of course is an assumption that will hold in the large. If one is looking at the spread of an infection over a large geographical area it is quite a reasonable one that we would expect to be well substantiated by the data. Formally, distance is for this model the common cause rendering the probabilities of individuals' becoming infected conditionally independent.

From the standpoint of the philosophical discussion of causality most of us are unhappy with simply thinking of distance as a cause. On the other hand, matters are rather subtle here. We certainly think of the force of gravity as being inversely proportional to the square of the distance. In fact, a whole class of models of classical mechanics are concerned with the study of forces that are a function of distance only. So it is clear that this epidemic model that uses distance as a causal variable is in the great mechanical tradition as well. On second thought, probably most of us can easily absorb the usage of saying something like the cause is weaker because of the distance from the object. We still think of some object as being the cause, and not the distance, but the power of the cause is affected directly by the distance.

At bottom here is the traditional problem of action at a distance. Substantively, distance like time is never believed to be in and of itself a cause. Models in which distance becomes a cause are recognized by all and sundry as purely formal in character. Moreover, most of us have faith that such models can be expanded to show that distance as such is a spurious cause.

The deeper mystery is what to make of action at a distance, but it is beyond the scope of this survey to enter into this tangled problem. (I have done so previously in Suppes 1954; 1970, pp. 84–88.) The one point to remark on in the present context is that we certainly cannot give an adequate causal account of gravitational forces in the sense of classical physics if we have as an apparatus only the probability of discrete events. Fortunately, even ordinary talk is saturated with the qualitative idea of a continuously varying quantity. For example, *As he came nearer, his expectation of success increased*. An introductory probabilistic causal account of quantitative variables is given in my 1970 monograph, but a more thorough development in relation to substantive physical theories is needed.

Medical example. A standard approach to diagnosis and treatment in medicine is to assume that there exists a set of symptoms that serve as "first level" common causes for the sickness. In this case, the symptoms are highly restricted in character, for example, a high temperature, diarrhea, etc. At the next level we look for "real" common causes. For example, we may find the presence of some bacterium or some virus. But this identification of "real" common causes – being a bacterium, for example – is not one that is the least bit satisfactory from the standpoint of fundamental molecular biology. One wants to understand the molecular action of the bacteria: exactly how do the bacteria cause collections of cells to show phenomenological symptoms. Notice in this case there is no real question of spatial distance. We very much think of the action of the bacteria in the cell as being by ordinary scale a case of contiguous action, not action at a distance.

Classical statistical example. A classical example of common causes thought of in terms of a normal distribution at a fixed time and therefore, in the general sense we are using here, a spatial distribution, is this. Let us restrict ourselves for purposes of notation to two phenomenological random variables X and Y, which have a bivariate normal distribution. We search for a common cause Z under which the phenomenological correlation of X and Y become zero, i.e., the partial correlation of X and Y with Z held constant is zero. To construct such a common cause Z we proceed as follows:

Set

(1) $E(Z) = 0,\ E(Z^2) = 1.$

(2) If $\rho(X, Y) \geqslant 0$, set

$$\rho(X, Z) = \rho(Y, Z) = \sqrt{\rho(X, Y)}.$$

(3) If $\rho(X, Y) < 0$, set

$$\rho(X, Z) = -\rho(Y, Z) = \sqrt{|\rho(X, Y)|}.$$

Then X, Y, and Z have a multivariate normal distribution with

$$\rho(XY \cdot Z) = 0,$$

where the period in the middle of the line is used to denote partial correlation with Z held constant.

2. TIME

At a fundamental philosophical and scientific level, probably none of us believes in temporal action at a distance. In principle, the structure of an object or of a phenomenon at a given time completely determines what will happen next, including the laws governing inherent randomness. At a fundamental level, we might well claim that direct causes never recede backward in time. History, from this causal perspective, is a trivial subject because of the absence of knowledge of structure. Put another way, this philosophical claim is an expression of the deeply entrenched view that phenomena, at the fundamental level, are entirely Markovian in character. This same assumption certainly dominates the general theoretical ideas of classical physics.

On the other hand, we all recognize the scientific hopelessness of such a position. No one would begin to suggest that we could in fact eliminate the historical view as the proper way to study a wide variety of phenomena. I am not talking simply about history proper here but about a wide range of disciplines from meteorology to economics. For example, if I want to predict what the annual rainfall will be for the coming year in a given spot, it is out of the question to examine only the current state of the atmosphere to make the prediction. The only chance of coming reasonably close is to look at data from past years, possibly modified with consideration of a few current local parameters. But certainly the dominant term in the prediction will be the data from the past.

Grand example. The marvelous success story of the view that current structure determines current properties and is accessible at least indirectly to determination is the theory of atomic structure. To a very large extent, the atomic structure of many objects determines in a way open to investigation their main physical properties independent of the history of the objects. One might say that what I have just said is not true of very many natural objects. It is certainly true of a variety of "pure" objects of great importance in the modern development of physical theory and also in modern engineering. The example of the theory of atomic structure is a great *scientific* success story. Philosophically, we might say it is not of great general significance because we all believe anyhow that current structure determines present and future behavior, but there are in fact few real cases that are computationally feasible. Yet those few cases have turned out to be of such importance scientifically and technologically that they constitute collectively the most surprising results of the last three centuries. Notice without being finicky how we think about causes in this instance. We think directly of the structure of the atoms as determining the present and future behavior of objects – as being the common cause of observable properties. The many subtle and intricate results of current solid state physics show very well that the theory of atomic structure is a subject that is very much alive and well and will be as important to the future as it has been in the past.

Chains of infinite order. Let us return to those many phenomena that cannot be given a working scientific analysis in terms of their current structure. A general framework for treating such examples is as a chain of infinite order. Put in probabilistic terms, suppose we have an infinite sequence of random variables $X_1, \ldots, X_n, \ldots$, with in general the outcome of random variable X_n being dependent on the outcomes of the preceding $n - 1$ random variables. We have in mind here the sequence unfolding in time, and we could just as well use a continuous-time index as the discrete one. (The notation will be somewhat simpler if time is indexed by discrete trials rather than continuously; I also omit the generalization to an infinite past.) What we often aim at scientifically is to obtain a Markov truncation of a chain of infinite order by introducing a concept of current state with the standard common cause property: once we conditionalize on this current state no further knowledge about the past will be of any use.

Unfortunately, however, unless we put some restrictions as in the earlier case we can trivialize the problem of finding a Markov truncation. Let us write the probability of response j on trial $n + 1$ conditional on the preceding n outcomes in the following fashion:

$$P(X_{n+1} = j | x_n, \ldots, x_1) = P(X_{n+1} = j | [x_n]),$$

where $[x_n]$ is a standard notation for the sequence of actual responses up to trial n. Obviously we get a Markov truncation by simply introducing at each stage states that are isomorphic to the sequence of first n responses, i.e., the Markov truncating state at trial $n = [x_n]$. Notice how uninteresting this result is, even more so than Theorem 1 about common causes, for here states are unbounded in number and have no natural intrinsic structure. We just keep introducing new states at each trial as we need them. The scientific objective, of course, is to do something very different (for more detailed discussion, see Suppes 1986).

In certain situations a change in random variables can make a chain of infinite order a Markov chain. For example, a model of learning or adaptive control that is linear in response probabilities can take as new random variables the probabilities of responses, rather than the responses themselves. One of the early examples of the effective use of this strategy is Karlin's (1953) derivation of detailed limit theorems for certain processes. More recently this approach has also been extensively used by Norman (1972). On the other hand, as already pointed out by Lamperti and Suppes (1959), this strategy will not work when probabilities of correction or adjustment depend on responses on previous trials. In such instances there is no apparent effective way to reduce the chains of infinite order to a Markov chain by introducing new random variables.

Stochastic processes. The widespread use of stochastic processes of all sorts, Markovian or not, is testimony to the great importance in current science of the intensive study of probabilistic theories of causality with explicit emphasis on temporal development of the causal processes under study. There is one important point of a general philosophical nature I want to make about such processes. A great variety of processes that are considered of central importance scientifically are extraordinarily complicated in detail. A familiar classic example is that of Brownian motion. For many kinds of processes of a simpler nature we are interested in studying the detailed sample paths of the particles or other objects in the process. What is important, however, about complicated processes like those of Brownian motion is the hopeless character of a detailed study of the actual sample paths. Imagine for a moment the totally unmanageable complexity of analyzing the actual paths of a few billion particles in a small sample of a gas – even if we could overcome the practical problems of making the observations we would not really know what to do with all the details of the actual paths once we had recorded them. This means that the theory of probabilistic causality often comes especially into its own when we are not looking at detailed sample paths but looking at various probabilistic features of some complicated phenomena. Brownian motion is an important example from classical physics with applications in many other domains, including quantum physics. An example of a very different sort is modeling something complicated like the activity of a telephone exchange. Here also the interest is not really in the details of each call but rather a good understanding of the dynamic behavior of the distribution of calls.

As perhaps is evident, the sermon I want to preach on this point is that it is a conceptual mistake to think we are going to ever understand the details of many kinds of complex phenomena. One of the important kinds of applications of probabilistic causality in actual scientific work, both theoretically and experimentally, is to the analysis of such complex phenomena. The important point is the identification of those random variables which represent aggregate behavior in some form and which can be studied in a meaningful fashion.

3. SPACE-TIME

As remarked earlier, I restrict myself here to special relativity only, because I know of scarcely any detailed discussions of probabilistic causality in the context of general relativity – there are few enough combining both probability and special relativity.

First, it can be observed that the general theory of probabilistic causality goes through in special relativity with the restriction that temporally earlier causes must lie in the back light-cone. In Figure 1 we are considering event A_t. The back light-cone of A_t is what is shown

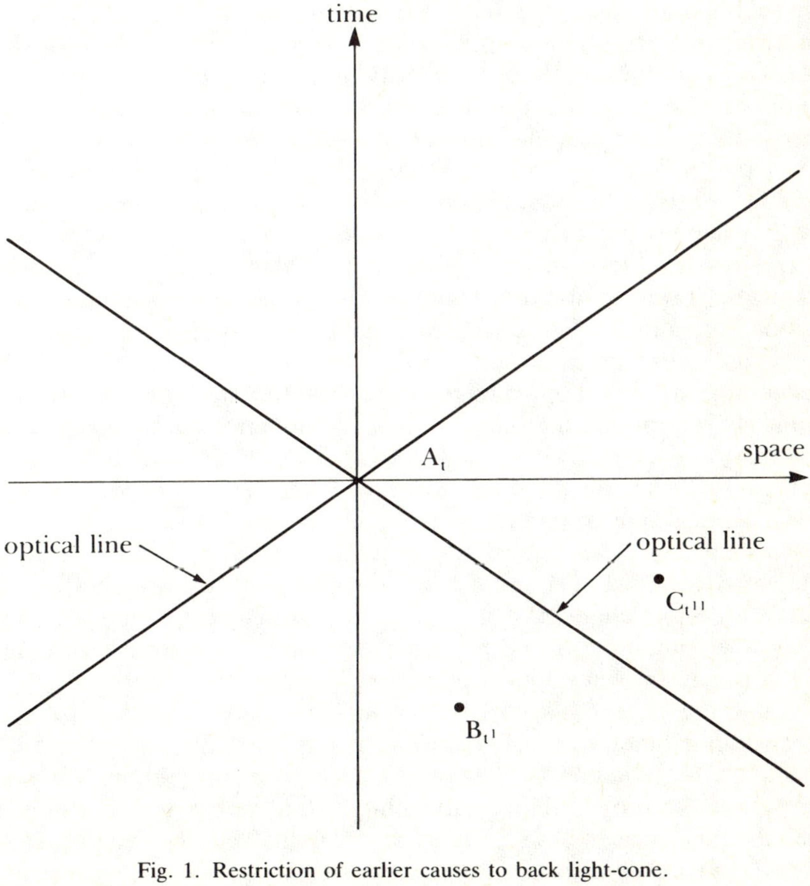

Fig. 1. Restriction of earlier causes to back light-cone.

as the open triangular region immediately below A_t bounded by light lines on each side in this one-dimensional spatial case, with the horizontal axis representing space and the vertical axis representing time. Note that the event $B_{t'}$ can be a cause of A_t because it lies in the back light-cone, but on the other hand, $C_{t''}$ cannot because it lies outside the back light-cone. This restriction of special relativity on causal action is one that we could easily impose on probabilistic causality and represents in principle only a slight modification in the general theory.

As long as we are dealing with the probability of point events and just looking at the conditional probability of one event on another, for example, no new fundamental problems are created, but it is quite another matter when we turn to what are meant to correspond to continuous distributions of properties in space. Note that in a process of classical physics, at time t a quantity may well be distributed over all of space and what we are studying is the dynamic change of that distribution with time. The immediate question for probabilistic causality in the framework of special relativity is, what is the proper domain of such a continuous distribution. (I have discussed these matters earlier (Suppes 1981), but I want to expand here on what I had to say.)

Let us take a look first at the one-dimensional case. So suppose we have a univariate normal distribution. In the classical case that distribution has as its domain simply the one dimension of space. In special relativity we would naturally pick one of two types of lines, either a separation line or an optical line (by separation line I mean a line that connects two points that cannot causally influence each other). Such lines are also called spatial lines. Already even in the one-dimensional case, if we pick optical lines we have immediately the question of which optical line through a given point, as shown in Figure 2.

This concept of an optical line generalizes to an optical plane or an optical threefold. Let A and B be two optical lines that are parallel and neutral, i.e., no element of A is before or after any element of B. Then the plane containing A and B is an *optical* plane. We generalize in the obvious way from an optical plane to an optical threefold.

But there is, I think, a better solution – closer to the space of Euclidean geometry and classical mechanics. First, it will be desirable to sketch the geometry of classical, i.e., Galilean, space-time. We need to assume, or prove, among other things, that we have a four-dimensional affine space, based, let us say, on the ternary relation B of betweenness. We also need a concept of simultaneity $\sim$, which is a binary relation between point events, and as well a quaternary relation

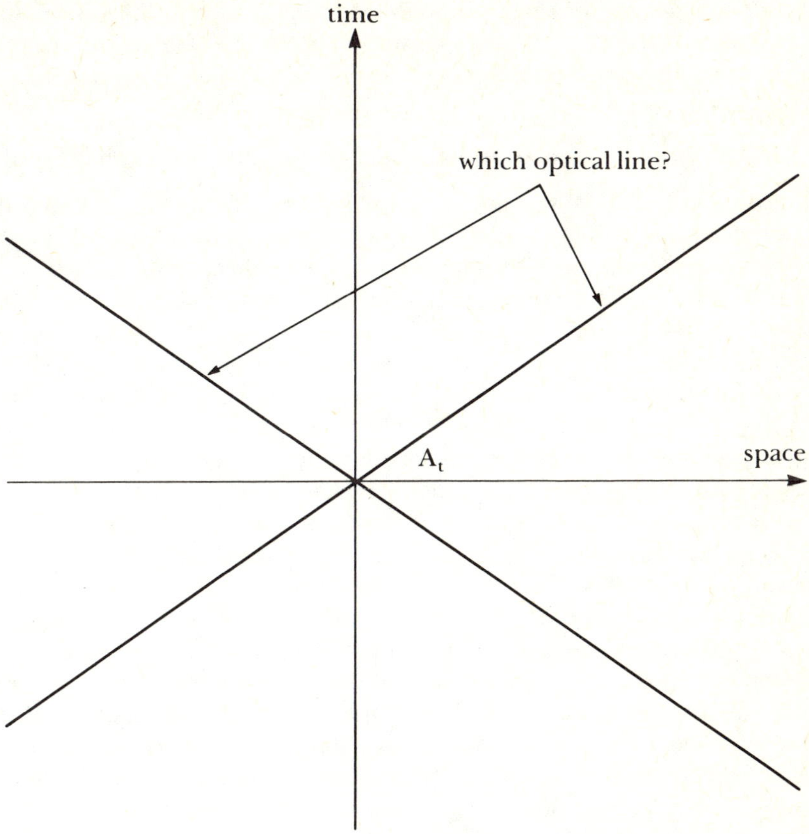

Fig. 2. Which optical line to choose as domain of probability distribution.

$\approx$ of equidistance or congruence. Let A be the set of all (point) events. Then for any event a we define the set of events simultaneous with a: $[a] = \{b : b \in A \;\&\; b \sim a\}$. Then, an axiom or a theorem is that the structure $([a], B, \approx)$ for each event a is a three-dimensional Euclidean geometry. In the classical case, probability distributions are defined on these spaces, and the stochastic problem is to understand their dynamics in time.

In the relativistic case we have, of course, no direct concept of simultaneity of distant events, but we do have something to replace it. Some preliminary definitions are needed. An *optical parallelogram* is a parallelogram whose sides are segments of optical lines. An inertial line and a spatial line are *orthogonal* if and only if they form the diagonals of an optical parallelogram. Now, in the relativistic case we select not only a point event a but also an inertial line L through a. We then define:

$$[a]_L = \{b : b \in A \ \& \ ab \text{ is a spatial line orthogonal to } L\}.$$

Note that if $b \in [a]_L$ the line ab must intersect L at a. The axiom or theorem of the geometry of Minkowski space we now need is this. *For every event a and inertial line L through a, the structure* $([a]_L, B, \approx)$ *is a three-dimensional Euclidean space*. Gaussian or normal probability distributions, so familiar in physics, have then such a standard spatial domain, even in the relativistic case.

Of course, we could pick as the domain the forward or back light-cone of a point, but then the standard affine properties of the normal distribution are lost. The spatial choice indicated seems to be the right one for physical purposes, at least for mean distributions.

The reservation just indicated is important. There are difficulties in extending in any direct way the full theory of Brownian motion to special relativity (for a review of the problems, see Nelson, 1978). But the causal theory of classical quantum mechanics is not so detailed as that of classical Brownian motion. Particles do not have well-defined trajectories. From a probabilistic standpoint what we mainly get are mean distributions, and the Schroedinger equation provides the causal theory of how the *mean* distribution changes under the action of a given Hamiltonian or potential. Relativistic quantum mechanics shares these same features. The Dirac theory of the election is a standard specific example in which a probability density defined over the whole of space as characterized above arises in a natural way along with the probability current.

It is essential to conserve as long as possible in any theoretical developments the affine spaces that support the multivariate normal distribution. Detailed dynamic, i.e., causal, theories for any other family of continuous multivariate distributions scarcely exist. Holding on to ordinary Euclidean space within the framework of special relativity remains a critical matter for probabilistic causality as expressed in relativistic quantum mechanics, even if such distributions do not have all the invariant features we would like.

Stanford University

REFERENCES

Bell, J. S. (1964) 'On the Einstein–Podolsky–Rosen paradox', *Physics* **1**, 195–200.

Clauser, J. F., Horne, M. A., Shimony, A., and Holt, R. A. (1969) 'Proposed experiment to test local hidden-variable theories', *Physical Review Letters* **23**, 880–884.

Cressie, N. and Holland, P. W. (1983) 'Characterizing the manifest probabilities of latent trait models', *Psychometrika* **48**, 129–141.

Fine, A. (1982) 'Hidden variables, joint probability, and the Bell inequalities', *Physical Review Letters* **48**, 291–295.

Garg, A. and Mermin, N. D. (1982) 'Correlation inequalities and hidden variables', *Physical Review Letters* **49**, 1220–1223.

Holland, P. W. (1981) 'When are item response models consistent with observed data?' *Psychometrika* **46**, 79–92.

Holland, P. W. and Rosenbaum, P. R. (1986) 'Conditional association and unidimensionality in monotone latent variable models', *Annals of Statistics* **14**, 1523–1543.

Holland, P. W. and Thayer, D. T. (1985) 'Section pre-equating in the presence of practice effects', *Journal of Educational Statistics* **10**, 109–120.

Karlin, S. (1953) 'Some random walks arising in learning models 1', *Pacific Journal of Mathematics* **3**, 725–726.

Lamperti, J. and Suppes, P. (1959) 'Chains of infinite order and their application to learning theory', *Pacific Journal of Mathematics* **9**, 739–754.

Nelson, E. (1978) 'Le mouvement Brownien relativiste', by Jean-Pierre Caubet, *Lecture Notes in Mathematics* **559** (Springer-Verlag, Berlin, Heidelberg, New York, 1976.) Book review, *Bulletin of the American Mathematical Society* **84**, 121–124.

Norman, M. F. (1972) *Markov Processes and Learning Models.* (New York: Academic Press.)

Rosenbaum, P. R. (1984) 'Testing the conditional independence and monotonicity assumptions of item response theory', *Psychometrika* **49**, 425–435.

Shimony, A. (1981) 'Critique of the papers of Fine and Suppes', in P. D. Asquith and R. N. Giere (eds.), *PSA 1980* (Vol. 2). (East Lansing, Mich.: Philosophy of Science Association.) Pp. 572–580.

Suppes, P. (1954) 'Descartes and the problem of action at a distance', *Journal of the History of Ideas* **15**, 146–152.

Suppes, P. (1970) *A Probabilistic Theory of Causality.* (*Acta Philosophica Fennica* **24**) (Amsterdam: North-Holland.)

Suppes, P. (1981) 'Probability in relativistic particle theory', *Erkenntnis* **16**, 299–305.

Suppes, P. (1986) 'Non-Markovian causality in the social sciences with some theorems on transitivity', *Synthese* **68**, 129–140.

Suppes, P. and Zanotti, M. (1980) 'A new proof of the impossibility of hidden variables using the principles of exchangeability and identity of conditional distribution', in P. Suppes (ed.), *Studies in the Foundations of Quantum Mechanics.* (East Lansing, Mich.: Philosophy of Science Association.) Pp. 173–191.

Suppes, P. and Zanotti, M. (1981) 'When are probabilistic explanations possible?' *Synthese* **48**, 191–199.

Suppes, P. and Zanotti, M. (to appear). 'Existence of joint distribution with given covariances and Bell's inequalities in quantum mechanics'.

PART B

PHYSICAL PROBABILITY, DEGREE OF BELIEF, AND De FINETTI'S THEOREM

S. L. ZABELL

SYMMETRY AND ITS DISCONTENTS

The following paper consists of two parts. In the first it is argued that Bruno de Finetti's theory of subjective probability provides a partial resolution of Hume's problem of induction, if that problem is cast in a certain way. De Finetti's solution depends in a crucial way, however, on a symmetry assumption – exchangeability – and in the second half of the paper the broader question of the use of symmetry arguments in probability is analyzed. The problems and difficulties that can arise are explicated through historical examples which illustrate how symmetry arguments have played a important role in probability theory throughout its development. In a concluding section the proper role of such arguments is discussed.

1. THE DE FINETTI REPRESENTATION THEOREM

Let X_1, X_2, X_3, ... be an infinite sequence of 0,1-valued random variables, which may be thought of as recording when an event occurs in a sequence of repeated trials (e.g., tossing a coin, with 1 if heads, 0 if tails). The sequence is said to be *exchangeable* if all finite sequences of the same length with the same number of ones have the same probability, i.e., if for all positive integers n and permutations σ of $\{1, 2, 3, \ldots, n\}$,

$$P[X_1 = e_1, X_2 = e_2, \ldots, X_n = e_n] = P[X_1 = e_{\sigma(1)}, X_2 = e_{\sigma(2)}, \ldots, X_n = e_{\sigma(n)}],$$

where e_i denotes either a 0 or a 1. For example, when $n = 3$, this means that

$$P[1, 0, 0] = P[0, 1, 0] = P[0, 0, 1] \quad \text{and} \quad P[1, 1, 0] = P[1, 0, 1] = P[0, 1, 1].$$

(Note, however, that $P[1, 0, 0]$ is not assumed to equal $P[1, 1, 0]$; in general, these probabilities may be quite different.)

In 1931 the Italian probabilist Bruno de Finetti proved his famous

Brian Skyrms and William L. Harper (eds.), Causation, Chance, and Credence, Vol. I, 155–190.

DE FINETTI REPRESENTATION THEOREM. *Let $X_1, X_2, X_3, \ldots$ be an infinite exchangeable sequence of* 0,1-*valued random variables, and let $S_n = X_1 + X_2 + \cdots + X_n$ denote the number of ones in a sequence of length n. Then it follows that:*

1. *the limiting frequency $Z =: \lim_{n\to\infty}(S_n/n)$ exists with probability 1.*
2. *if $\mu(A) =: P[Z \in A]$ is the probability distribution of Z, then*

$$P[S_n = k] = \int_0^1 \binom{n}{k} p^k (1 - p)^{n-k} \, d\mu(p)$$

for all n and k.[1]

This remarkable result has several important implications. First, contrary to popular belief, subjectivists clearly believe in the existence of infinite limiting relative frequencies – at least to the extent that they are willing to talk about an (admittedly hypothetical) infinite sequence of trials.[2] The existence of such limiting frequencies follows as a purely mathematical consequence of the assumption of exchangeability.[3] When an extreme subjectivist such as de Finetti denies the existence of objective chance or physical probability, what is really being disputed is whether limiting frequencies are objective or physical properties.

There are several grounds for such a position, but all center around the question of what "object" an objective probability is a property of. Surely not the infinite sequence, for that is merely a convenient fiction (Jeffrey 1977). Currently the most fashionable stance seems to be that objective probabilities are a *dispositional property* or *propensity* which manifests itself in, and may be measured with ever-increasing accuracy by, finite sequences of ever-increasing length (see, e.g., Kyburg 1974).

But again, a property of what? Not the coin, inasmuch as some people can toss a so-called "fair" coin so that it lands heads 60% of the time or even more (provided the coin lands on a soft surface such as sand rather than a hard surface where it can bounce). Some philosophers attempt to evade this type of difficulty by ascribing propensities to a *chance set-up* (e.g., Hacking 1965): in the case of coin-tossing, the coin *and* the manner in which it is tossed. But if the coin were indeed tossed in an identical manner on every trial, it would always come up heads or always come up tails; it is precisely because the manner in which the coin is tossed on each trial is *not* identical that the coin can come up both ways. The suggested chance set-up is in fact nothing other than a sequence of objectively differing trials which we are subjectively unable to distinguish between. At best, the infinite limiting frequency is a property of an "object" enjoying both objective and subjective features.

2. DE FINETTI VANQUISHES HUME

The most important philosophical consequence of the de Finetti representation theorem is that it leads to a solution to *Hume's problem of induction:* why should one expect the future to resemble the past? In the coin-tossing situation, this reduces to: in a long sequence of tosses, if a coin comes up heads with a certain frequency, why are we justified in believing that in future tosses of the same coin, it will again come up heads (approximately) the same fraction of the time?

De Finetti's answer to this question is remarkably simple. Given the information that in n tosses a coin came up heads k times, such data is incorporated into one's probability function via

Bayes's rule of conditioning: $P[A \mid B] = P[A \text{ and } B]/P[B]$.

If n is large and $p^* = k/n$, then – except for certain highly opinionated, eccentric, or downright kinky "priors" $d\mu$ – it is easy to prove that the resulting posterior probability distribution on p will be highly peaked about p^*; that is, the resulting probability distribution for the sequence of coin tosses looks approximately like (in a sense that can be made mathematically precise) a sequence of independent and identically distributed Bernoulli trials with parameter p^* (i.e., independent tosses of a p^* coin). By the weak law of large numbers it follows that, with high probability, subsequent tosses of the coin will result in a relative frequency of heads very close to p^*.

Let us critically examine this argument. Mathematically it is, of course, unassailable. It implicitly contains, however, several key suppositions:

1. P is operationally defined in terms of betting odds.
2. P satisfies the axioms of mathematical probability.
3. P is modified upon the receipt of new information by Bayesian conditioning.
4. P is assumed to be exchangeable.

In de Finetti's system, degree of belief is quantified by the betting odds one assigns to an event. By a Dutch book or coherence argument, one deduces that these betting odds should be consistent with the axioms of mathematical probability. Conditional probabilities are initially defined in terms of conditional bets and Bayes's rule of conditioning is deduced as a consequence of coherence. The relevance of conditional probabilities to inductive inference is the *dynamic assumption of Bayesianism* (Hacking 1967): if one learns that B has occurred, then one's new probability assignment is $P[A \mid B]$. In general, however, conditional probabilities can behave in very nonHumeian ways, and (infinite) exchangeability is taken as describing the special class of situations in which Humeian induction is appropriate.

This paper will largely concern itself with the validity of this last assumption. Suffice it to say that, like Ramsey (1926), one may view the subjectivist interpretation as simply capturing *one* of the many possible meanings or useages of probability; that the Dutch book and other derivations of the axioms may be regarded as plausibility arguments (rather than normatively compelling); and that although a substantial literature has emerged in recent decades concerning the limitations of Bayesian conditioning, the difficulties discussed and limitations raised in that literature do not seem particularly applicable to most of the situations typically envisaged in discussions of Hume's problem.

The assumption of exchangeability, however, seems more immediately vulnerable. Isn't it essentially circular, in effect assuming what one wishes to prove? Of course, in one sense this must obviously be the case. All mathematics is essentially tautologous, and any implication is contained in its premises. Nevertheless, mathematics has its uses. Formal logic and subjective probability are both theories of consistency, enabling us to translate certain assumptions into others more readily palatable.

What de Finetti's argument really comes down to is this: if future outcomes are viewed as exchangeable, i.e., no one pattern is viewed as any more or less likely than any other (with the same number of successes), then when an event occurs with a certain frequency in an initial segment of the future, we must, if we are to be consistent, think it likely that that event will occur with approximately the same frequency in later trials. Conversely, if we do not accept this, it means that we must have – prospectively – thought certain patterns more likely than others. Which means that we must have possessed more information than is ordinarily posited in discussions of Humeian induction.

And there the matter would appear to stand. Or does it?

3. THE INSIDIOUS ASSUMPTION OF SYMMETRY

Exchangeability is one of many instances of the use of symmetry arguments to be found throughout the historical development of mathematical probability and inductive logic. But while such arguments often have a seductive attraction, they also often carry with them "hidden baggage": implications or consequences, sometimes far from obvious, which later cast serious doubt on their validity. We will discuss three historically important examples, all involving attempts to justify induction by the use of probability theory, and all (in effect) involving the appropriate choice of prior $d\mu$ in the de Finetti representation.

EXAMPLE 3.1: *Bayes's argument for the Bayes–Laplace prior.*

Consider "an event concerning the probability of which we absolutely know nothing antecedently to any trials made concerning it" (Bayes 1764). Implicitly invoking a symmetry argument, Bayes argued that "concerning such an event I have no reason to think that, in a certain number of trials, it should rather happen any one possible number of times than another," i.e., that in a sequence of n trials one's probability assignment for S_n, the number of heads, should satisfy

Bayes's Postulate: $P[S_n = k] = 1/(n + 1)$.

That is, the number of heads can assume any of the $n + 1$ values 0, 1, 2,. . ., n and, absent further information, all $n + 1$ values are viewed as equally likely. In a famous Scholium, Bayes concluded that if this were indeed the case, then the prior probability $d\mu(p)$ must be the "flat" prior dp.[4]

Although Bayes's exact reasoning at this point is somewhat unclear, it can easily be made rigorous: Taking $k = 0$ in the de Finetti representation and using Bayes's postulate, it follows that

$$\int_0^1 p^n \, d\mu(p) = 1/(n + 1).$$

The integral on the left-hand side is the n-th moment of $d\mu$, so Bayes's assumption uniquely determines the moments of $d\mu$. But since $d\mu$ is concentrated on a compact set, it follows by a theorem of Hausdorff that $d\mu$, if it exists, is in turn determined by its moments. That is, there can be at most one probability measure $d\mu$ which satisfies Bayes's assumption $P[S_n = k] = 1/(n + 1)$. But the flat measure dp does satisfy this integral equation, i.e.,

$$\int_0^1 p^n \, dp = 1/(n + 1),$$

hence $d\mu$ must be dp.

Bayes's argument is quite attractive. A modern-day subjectivist might view Bayes's assumption as a *definition* (possibly one of many) of "complete ignorance" (rather than consider "complete ignorance" to be an *a priori* meaningful concept), but would probably find Bayes's argument otherwise unobjectionable.

The argument in its original form, however, did not go uncriticized. As Boole (1854, pp. 369–375) noted, rather than consider the events $[S_n = k]$ to be equally likely, one could equally plausibly take all sequences of a fixed length (or "constitutions") to be so. Thus, for $n = 3$

$$\begin{aligned} P[000] &= P[100] = P[010] = P[001] = P[110] \\ &= P[101] = P[011] = P[111] = 1/8. \end{aligned}$$

To many, this assignment seemed a far more natural way of quantifying ignorance than Bayes's.

Unfortunately, it contains a time-bomb with a very short fuse. As Carnap (1950, p. 565) later noted (and Boole himself had already remarked), this probability assignment corresponds to *independent* trials, and thus remains unchanged when conditioned on the past, an obviously unsatisfactory choice for modeling inductive inference, inasmuch as "past experience does not in this case affect future expectation" (Boole 1854, p. 372).

In his *Logical Foundations of Probability* (1950), Carnap announced that in a later volume, "a quantitative system of inductive logic" would be constructed, based upon a function Carnap denoted c^*. Carnap's c^* function was, in effect, the one already proposed by Bayes. But

Carnap grew uneasy with this unique choice, and in his monograph *The Continuum of Inductive Methods* (1952), he advocated instead the use of a one-parameter family containing c^*. Unknown to Carnap, however, he had been anticipated in this, almost a quarter of a century earlier, by the English philosopher William Ernest Johnson.

EXAMPLE 3.2: *W. E. Johnson's sufficientness postulate.*

In 1924 Johnson, a Cambridge logician, proposed a multinomial generalization of Bayes's postulate. Suppose there are $t \geqslant 2$ categories or types, and in n trials there are n_1 outcomes of the first type, n_2 outcomes of the second type, . . . , and n_t outcomes of the t-th type, so that $n = n_1 + n_2 + \cdots + n_t$. The sequence $(n_1, n_2, \ldots, n_t)$ is termed an *ordered t-partition of n*. Bayes had considered the case $t = 2$, and his postulate is equivalent to assuming that all ordered 2-partitions $(k, n - k)$ are equally likely. Now Johnson proposed as its generalization

Johnson's combination postulate: Every ordered t-partition of n is equally likely.

For example, if $t = 3$ and $n = 4$, then there are 15 possible ordered 3-partitions of 4, viz.:

n_1	n_2	n_3
4	0	0
3	1	0
3	0	1
2	2	0
2	1	1
2	0	2
1	3	0
1	2	1
1	1	2
1	0	3
0	4	0
0	3	1
0	2	2
0	1	3
0	0	4

and each of these is assumed to be equally likely.

Johnson did not work with integral representations but, like Carnap, with finite sequences. In so doing he introduced a second postulate, his "permutation postulate." This was none other the assumption of exchangeability, thus anticipating de Finetti (1931) by almost a decade! (If one labels the types or categories with the letters of a t-letter alphabet, exchangeability here means that all words of the same length, containing the same number of letters of each type, are equally likely). Together, the combination and permutation postulates uniquely determine the probability of any specific finite sequence. For example, if one considers the fifth partition in the table above, $4 = 2 + 1 + 1$, then there are twelve sequences which give rise to such a partition, viz.

x_1	x_2	x_3	x_4
1	1	2	3
1	1	3	2
1	2	1	3
1	2	3	1
1	3	1	2
1	3	2	1
2	1	1	3
2	1	3	1
2	3	1	1
3	1	1	2
3	1	2	1
3	2	1	1

and each of these are thus assumed to have probability $(1/15)(1/12) = 1/180$. The resulting probability assignment on finite sequences is identical with Carnap's c^*.

Despite its mathematical elegance, Johnson's "combination postulate" is obviously arbitrary, and Johnson was later led to substitute for it another, more plausible one, his "sufficientness postulate." This new postulate assumes for all n

Johnson's sufficientness postulate:

$$P[X_{n+1} = j \mid X_1 = i_1, X_2 = i_2, \ldots, X_n = i_n] = f(n_j, n).$$

That is, the conditional probability that the next outcome is of type j depends only on the number of previous trials and the number of previous outcomes of type j, but not on the frequencies of the other types or the specific trials on which they occurred. If, for example $t = 3$, $n = 10$, and $n_1 = 4$, the postulate asserts that on trial 11 the (conditional) probability of obtaining a 1 is the same for all sequences containing four 1's and 6 not -1's, and that this conditional probability does not depend on whether there were six 2's and no 3's, or five 2's and one 3, and so on. (Note that the postulate implicitly assumes that all finite sequences have positive probability, so that the conditional probabilities are well-defined.)

Johnson's sufficientness postulate makes what seems a minimal assumption: absence of knowledge about different types is interpreted to mean that information about the frequency of one type conveys no information about the likelihood of other types occurring. It is therefore rather surprising that it follows from the postulate that the probability function P is uniquely determined up to a constant:

THEOREM. (Johnson 1932): *If P satisfies the sufficientness postulate and $t \geqslant 3$, then there exists a $k > 0$ such that*

$$f(n_i, n) = \{n_i + k\}/\{n + tk\}.$$

This is, of course, nothing other than Carnap's "continuum of inductive methods."[5]

The de Finetti representation theorem can be generalized to a much wider class of infinite sequences of random variables than those taking on just two values (see, e.g., Hewitt and Savage 1955). In the multinomial case now being discussed, the de Finetti representation states that every exchangeable probability can be written as a mixture of multinomial probabilities. Just as Bayes's postulate implied that the prior $d\mu$ in the de Finetti representation was the flat prior, Johnson's theorem implies that the mixing measure $d\mu$ in the de Finetti representation is the *symmetric Dirichlet prior*

$$p_1^{k-1} p_2^{k-1} \ldots p_t^{k-1} \, dp_1 \, dp_2 \ldots dp_{t-1}:$$

a truly remarkable result, providing a subjectivistic justification for the use of the mathematically attractive Dirichlet prior.[6]

Despite its surface plausibility, Johnson's sufficientness postulate is often too strong an assumption. While engaged in cryptanalytic work for the British government at Bletchley Park during World War II, the English logician Alan Turing realized that even if one lacks specific knowledge about individual category types, the frequencies $n_1, n_2, \ldots, n_t$ may contain relevant information about predictive probabilities, namely the information contained in the *frequencies of the frequencies*.

Let a_r = the number of frequencies n_i equal to r; a_r is called the frequency of the frequency r. For example, if $t = 4$, $n = 10$, and one observes the sequence 4241121442, then $n_1 = 3$, $n_2 = 3$, $n_3 = 0$, $n_4 = 4$

and $a_0 = 1$, $a_1 = 0$, $a_2 = 0$, $a_3 = 2$, $a_4 = 1$. (A convenient shorthand for this is $0^1 1^0 2^0 3^2 4^1$.) Although it is far from obvious, the a_r may be used to estimate cell probabilities: see Good (1965, p. 68).[7]

EXAMPLE 3.3: *Exchangeability and Partial Exchangeability.*

Given the failure of such attempts, de Finetti's program must be seen as a further retreat from the program of attempting to provide a unique, quantitative account of induction. Just as Johnson's sufficientness postulate broadened the class of inductive probabilities from that generated by the Bayes–Laplace prior to the continuum generated by the symmetric Dirichlet priors, so de Finetti extended the class of possible inductive probabilities even further to include *any* exchangeable probability assignment.

But what of the symmetry assumption of exchangeability? Even this is not immune to criticism (as de Finetti himself recognized). Consider the following sequence: 000101001010100010101001.... Scrutiny of the sequence reveals the interesting feature that although every 0 is followed by a 0 or 1, every 1 is invariably followed by a 0. If this feature were observed to persist over a long segment of the sequence (or simply that 1's were followed by 0's with high frequency), then this would seem relevant information that should be taken into account when calculating conditional, predictive probabilities. Unfortunately, exchangeable probabilities are useless for such purposes: if P is exchangeable, then the conditional probabilities

$$P[X_{n+1} = j \mid X_1 = i_1, X_2 = i_2, \ldots, X_n = i_n]$$

depend solely on the number of 1's, and not on their order within the sequence. Thus, exchangeability, despite its plausibility, rules out a natural form of inductive inference and can only be considered valid when "order effects" are ruled out (as, for example, in coin-tossing).

An appropriate generalization of exchangeability that takes such order information into account is the concept of *Markov exchangeability:* all sequences with the same initial letter and the same transition counts (t_{ij} =: number of transitions from state i to state j) are assumed equally likely. Here too a de Finetti representation is possible (Diaconis and Freedman 1980b, 1980c): now one mixes on the possible transition matrices p_{ij}.

Once one has come this far, of course, it is easy to recognize that order effects of this type are merely one of many possible patterns that may be judged to provide useful information, each pattern requiring a corresponding generalization of exchangeability to incorporate the information it provides. To deal with such situations, de Finetti introduced in 1938 the notion of *partial exchangeability* (Diaconis and Freedman 1980c). Although partial exchangeability is an active field of current mathematical research still undergoing development (see, e.g., Diaconis and Freedman 1985), the general outline of the theory is clear: to each pattern corresponds a statistic or symmetry, a representation theorem, and a corresponding mode of inductive inference.

Thus, de Finetti's resolution of Hume's problem of induction is a highly qualified one: it is a theory of coherence. Every person's probability function will contain some symmetry involving past and future, and coherence dictates that patterns observed in the past will be expected to recur in the future.

Despite its highly qualified nature, the above analysis has an important payoff: it demonstrates that Hume's problem is in fact ill-posed; to ask "*why* should the future be expected to resemble the past?" presupposes having already answered the question "*how* is the future expected to resemble the past?" (It is essentially this point that is behind Nelson Goodman's "grue" paradox.) It is a strength of the subjectivist analysis that this point emerges as natural and obvious; indeed, it is essentially forced on one; and to the extent that one can state precisely the ways in which the past and future are conjectured to correspond, it gives a satisfactory solution to Hume's problem.

The successive attempts of Bayes, Johnson, and de Finetti to solve the problem of induction are marked by the invocation of progressively weaker symmetry assumptions. Symmetry, however, has played not only a key role in the the attempts to quantify induction, it has played a central role in the birth and evolution of probability theory, more central perhaps than sometime recognized. In the next three sections it will be argued that the birth of mathematical probability marked a key change in the way symmetry arguments were used; that the early dependence on symmetry arguments to quantify probability, while crucial to its mathematical development, blurred important epistemological distinctions; and that it was only with the challenging of precisely those symmetry arguments in the 19th century that the conceptual clarification of probability became possible.

4. OU MALLON

The simplest and oldest of such arguments is the use of physical or epistemic symmetry to identify a *fundamental probability set* or FPS, i.e., a partition of the space of possible outcomes into equiprobable alternatives. The recognition and use of such sets to compute numerical probabilities for complex events was a key step in the birth of mathematical probability. Once the ability to calculate probabilities in this simple case had been mastered, the outlines of the mathematical theory discerned, and its practical utility recognized, all else followed. Why were the mathematicians of the 17th century able to take this step, while the Greeks, despite their mathematical prowess and penchant for games of chance, were not? The crucial point to recognize is that while for the pioneers of the modern theory the equipossible elements of an FPS were *equally likely*, for the Greeks *none were possible*.

This was because of what G. E. L. Owen has described as "a very Greek form of argument" (Owen 1966), a form of reasoning employed by the Greeks that Leibniz was very fond of and which he called the *principle of sufficient reason*: "for every contingent fact there is a reason why the fact is so and not otherwise ..." (Broad 1975, p. 11). In the words of Leucippus (the only complete sentence of his which has come down to us), "Nothing occurs at random, but everything for a reason and by necessity" (Kirk and Raven 1957, p. 413). Two famous examples will illustrate its use:

4.1. *Anaximander and the position of the earth.* Anaximander (c. 610–540 B.C.), one of the early pre-Socratic Greek philosophers, believed the Earth lay at the center of the universe. But unlike Thales before him, who thought the Earth floated on water, and Anaximenes after, who thought it floated on air, Anaximander thought the Earth was unsupported and remained at the center for reasons of symmetry (*omoiotes*; variously translated as "similarity," "indifference," "equilibrium," or "equiformity").[8] Unfortunately, the text of Anaximander has not survived, and we are dependent on secondary, incomplete, and contradictory later accounts for information about the precise nature of his astronomical beliefs.[9] Our best source is perhaps Aristotle, who reports:

There are some who say, like Anaximander among the ancients, that [the earth] stays still because of its equilibrium. For it behoves that which is established at the center, and is equally related to the extremes, not to be borne one whit more either up or down or to the sides; and it is impossible for it to move simultaneously in opposite directions, so that it stays fixed by necessity. [*de Caelo* 295 b10]

How closely this reproduces Anaximander's own logic, the exact meaning to be attached to *omoiotes*, indeed the precise nature of the argument itself, is unclear. Nevertheless, the gist of the argument is clearly an appeal to symmetry: for every direction there is an opposite; since there is no more reason for the earth to move in one direction than another, the proper conclusion is that it moves in *neither*.

Although Aristotle expressed scepticism about such reasoning, it was fully accepted by Plato:

I am therefore persuaded that, in the first place, since the earth is round and in the middle of the heaven, it has not need either of air of any other necessity in order not to fall, but the similarity of the heaven to itself in every way and the equilibrium of the earth suffice to hold it still. For an equilibrated thing set in the midst of something of the same kind will have no reason to incline in one direction more than in another. But as its relationship is symmetrical it will remain unswervingly at rest. [*Phaedo* 108e–109a; c.f. *Timaeus* 62d.12]

4.2. *Parmenides and the creation of the universe.* Parmenides gave a similar argument to show that the universe had never been created:

And what need would have driven it on to grow, starting from nothing, at a later time rather than an earlier? [Kirk and Raven 1957, p. 273]

Again this is essentially a symmetry argument: if the universe had been created, it must have been at some specific time; inasmuch as there is no more reason for it to have been created at any one time than any other, all possible times are thereby ruled out. Obviously the argument makes some presuppositions, but it had great appeal to Leibniz and appears in his correspondence with Clarke.[10]

It is, as G. E. L. Owen notes,

a very Greek pattern of argument.... Aristotle retailored the argument to rebut the probability of motion in a vacuum; the Academy adapted it to show that, since no physical sample of equality has more right to serve as a standard sample than any other, the standard sample cannot be physical. And Leibniz found an excellent example in Archimedes's mechanics.... [Owen 1966]

The Greek Pyrrhonian skeptics made systematic use of a similar device for destroying belief. Their goal was to achieve a state of *epoche*, or suspension of judgement about statements concerning the external world, which they believed would in turn lead to *ataraxia*, a state of tranquility, "... saying concerning each individual thing that it no more [*ou mallon*] is than is not, or that it both is and is not, or that it neither is nor is not."[11]

How can *epoche* be achieved? According to Sextus Empiricus (*Outlines of Pyrrhonism* 1.8):

> Scepticism is an ability which sets up antitheses among appearances and judgments in any way whatever: by scepticism, on account of the 'equal weight' which characterizes opposing states of affairs and arguments, we arrive first at 'suspension of judgment', and second at 'freedom from disturbance'.

For example, knowledge of what is good is impossible, for what one person thinks good, another may think bad, and

> if we say that not all that anyone thinks good is good, we shall have to judge the different opinions; and this is impossible because of the equal validity of opposing arguments. Therefore the good by nature is impossible.

It is important to understand the implications of asserting "*ou mallon*." One might interpret it in a positive sense: although certain knowledge is ruled out, the information we possess is equally distributed between two or more possibilities, and hence we have an equal degree of belief in each. That this was *not* the skeptical position is clear from a passage in Diogenes Laertius (*Life of Pyrrho* 9.74–76):

> Thus by the expression "We determine nothing" is indicated their state of even balance; which is similarly indicated by the other expressions, "Not more (one thing than another)," "Every saying has its corresponding opposite," and the like. But "Not more (one thing than another)" can also be taken positively, indicating that two things are alike; for example, "The pirate is no more wicked than the liar." But the Sceptics meant it not positively but negatively, as when, in refuting an argument, one says, "Neither had more existence, Scylla or the Chimaera ..." Thus, as Timon says in the *Pytho*, the statement [*ou mallon*] means just absence of all determination and withholding of assent. The other statement, "Every saying, etc.," equally compels suspension of judgment; *when facts disagree, but the contradictory statements have exactly the same weight, ignorance of the truth is the necessary consequence.* [Emphasis added]

Pyrrhonian skepticism is an extreme position, and the later Academic skeptics developed a theory that combined skepticism about certain knowledge with a description of rational decision based on probable knowledge.[12] Under Carneades this theory included a scale of the varying degrees of conviction conveyed by an impression, depending on whether it was "credible," "credible and consistent," or "credible, consistent, and tested." Carneades's theory amounts to an early account of qualitative or comparative subjective probability, and one might expect that a later skeptic would go the final step and attempt to numerically measure or describe such degrees of conviction. That this did not happen, it may be argued, was a consequence of the *ou mallon* viewpoint. Witness Cicero's statement:

If a question be put to [the wise man] about duty or about a number of other matters in which practice has made him an expert, he would not reply in the same way as he would if questioned as to whether the number of the stars is even or odd, and say that he did not know; for in things uncertain there is nothing probable [*in incertis enim nihil est probabile*], but in things where there is probability the wise man will not be at a loss either what to do or what to answer. [Cicero *Academica* 2.110]

A 19th century enthusiast of the principle of insufficient reason would have little hesitation in assigning equal probabilities to the parity of the number of stars; this passage thus strikingly illustrates a chasm that had to be crossed before numerical probabilities could be assigned. Cicero was familiar with a theory of probability, indeed much of the *Academica* is devoted to a discussion of Academic probabilism and is one of our major sources of information about it. But for Cicero the probable was limited in its scope, limited in a way that precluded its quantification. The FPS was the basic setting for the early development of mathematical probability – but for Cicero it was a setting in which the very notion of probability itself was inapplicable.

Support for this thesis may be found in the writings of Nicole Oresme, the Renaissance astronomer and mathematician (ca. 1325–1382). Oresme discussed Cicero's example of the number of stars but, writing only a few centuries before the earliest known probability calculations, there is a clear difference:

The number of stars is even; the number of stars is odd. One of these statements is necessary, the other impossible. However, we have doubts as to which is necessary, so that we say of each that it is possible.... The number of stars is a cube. Now indeed, we say that it is possible but not, however, probable or credible or likely [*non tamen probabile aut opinabile aut verisimile*], since such numbers are much fewer than others.... The number of stars is not a cube. We say that it is possible, probable, and likely.... [Oresme 1966, p. 385]

To some, the revolutionary content of this passage lies in its quasi-numerical assertion of the improbability of the number of stars being a cube (due to the infrequency of cubic numbers). But its real novelty is Oresme's willingness to extend the realm of the probable. Having made that transition, the frequency-based assertions of probability and improbability he makes follow naturally.

Thus the key step in the birth of mathematical probability – the identification of fundamental probability sets in order to quantify probability – while seemingly so natural, in fact contains a major presupposition. The ancients used symmetry arguments to destroy belief, where we use them to quantify it. This "conceptual revolution" culminated in the 20th century statistical resort to physical randomization (e.g., in sampling, randomized clinical trials, and Monte Carlo simulations): the paradox of deliberately imposing disorder to acquire information. The uses of randomization throughout the ancient and medieval world, in contrast, although common and widespread (for example, in games of chance and fair allocation) all depended, in one way or another, solely on its property of loss of information.

But while the use of symmetry made the calculus of probabilities possible, it also contained the seeds of future confusion.

5. CHANCE AND EQUIPOSSIBILITY

The birth of probability was not an untroubled one. Probabilities are usually classified into two major categories – epistemic and aleatory – and a multitude of subcategories: propensities, frequencies, credibilities, betting odds, and so on. In settings where an FPS exists, all of these will usually have a common value, and the necessity of distinguishing among the different meanings is not a pressing one. But as the initial successes of the "doctrine of chances" spurred on its application to other spheres, this happy state of affairs ceased and the need for distinctions became inevitable.

Just what the proper domains of chance and probability were, however, remained unclear. For the calculus of probabilities was initially the "doctrine of chances," and paradoxically, while the Greeks failed to extend the realm of the probable to include fundamental probability sets, in the early days of the doctrine of chances some thought the notion of chance *only* applicable to such settings. A few examples will suggest the difficulties and confusions that occurred.

1. *Arbuthnot and the sex-ratio.* In 1711, Dr. John Arbuthnot, a

Scottish writer, physician to Queen Anne, and close friend of Swift and Pope, published a short paper in the *Philosophical Transactions of the Royal Society*, entitled 'An Argument for Divine Providence Taken From the Constant Regularity Observed in the Births of Both Sexes.' Using statistics from the London Bills of Mortality for the preceding 82 years, Arbuthnot observed that male births had exceeded female births in London for each year from 1629 to 1710. Noting that if male and female births were equally likely, the probability of such an outcome was extremely small (1 in 2^{82}), Arbuthnot rejected the hypothesis of equilikehood, making in effect the earliest known statistical test of significance. But Arbuthnot did not conclude that male and female births possessed unequal probabilities. Instead, he rejected outright the possibility that sex was due to chance, concluding that the excess of males was due to the intervention of divine providence; that "... it is Art, not Chance, that governs" (Arbuthnot 1711, p. 189).

In contrasting art with chance, Dr. Arbuthnot was merely displaying his classical erudition; the dichotomy between *techne* (art) and *tyche* (chance) being a commonplace of Greek philosophy.[13] What *is* new is his belief that chance is only operative when probabilities are equilikely; that otherwise some outside force must be acting, causing the imbalance, and that one could no longer refer to chance. His specific line of reasoning was quickly faulted by Nicholas Bernoulli: if sex is likened to tossing a 35-sided die, with 18 faces labelled "male," and 17 labelled "female," then Arbuthnot's data are entirely consistent with the outcome of chance.[14] This response to Arbuthnot's argument does not dispute that chance is limited to fundamental probability sets; it simply points out that more that one FPS is possible.

Arbuthnot's juxtaposition of chance and cause, and his belief that chances must be equal, is echoed in Hume. For Hume chance "properly speaking, is merely the negation of a cause":

> Since therefore an entire indifference is essential to chance, no one chance can possibly be superior to another, otherwise than as it is compos'd of a superior number of equal chances. For if we affirm that one chance can, after any other manner, be superior to another, we must at the same time affirm, than there is something, which gives it superiority, and determines the event rather to that side that the other: That is, in other words, we must allow of a cause, and destroy the supposition of chance; which we had before establish'd. A perfect and total indifference is essential to chance, and one total indifference can never in itself be either superior or inferior to another. This truth is not peculiar to my system, but is acknowledg'd by every one, that forms calculations concerning chances. [*Hume* 1739, p. 125]

Thus, for Hume, not merely the mathematical calculation of chances but the very existence of chance itself is dependent on an "entire," "perfect," and "total indifference" among the different possibilities. Was this "acknowledg'd by every one?" Examination of the works of Bernoulli, DeMoivre, and Laplace does not entirely bare out this claim. There the equality of chances appears as a mathematical device, not a metaphysical necessity. Nevertheless, the contrast of chance with "art," "design," or "cause," that "something, which gives it superiority," is a recurrent theme. De Moivre suggests that "we may imagine Chance and Design to be, as it were, in Competition with each other" (De Moivre 1756, p. v). "Chance" and "Design" here no longer means the presence and absence of a stochastic element, but a lack of uniformity in the probability distribution. Answering Nicholas Bernoulli, De Moivre says yes, Arbuthnot's birth data is consistent with an 18:17 ratio, but "this Ratio once discovered, and *manifestly serving to a wise purpose*, we conclude the Ratio itself, or if you will the *Form of the Die*, to be an Effect of *Intelligence and Design*" (De Moivre 1756, p. 253).

Uniformity in distribution was to be increasingly equated with absence of design or law, departure from uniformity with their presence. A famous example is Michell's argument in 1767 that optically double or multiple stars were physically so. Michell calculated that the observed clustering of stars in the heavens exceeded what could reasonably be expected if the stars were distributed at random (i.e., uniformly) over the celestial sphere, inferring "either design, or some general law" due to "the greatness of the odds against things having been in the present situation, if it was not owing to some such cause" (Michell 1767, p. 243). Michell's argument was the focus of debate for a brief period during the middle of the 19th century, a key issue being precisely this equation of uniformity with absence of law.[15]

The elements of a fundamental probability set enjoy this status for reasons which are both *aleatory* (i.e., physical or objective) and *epistemic*. The dichotomy between chance and design involves primarily the aleatory aspect of the FPS. Throughout the 18th century, the elements of an FPS were often defined in terms of equipossibility, a terminology which, as Hacking notes (1975, Chapter 14), permitted a blurring of the aleatory and epistemic aspects. The literature of the period furnishes many instances of this duality. In the *Ars Conjectandi*, for example, James Bernoulli refers to cases which are "equally possible, that is to say, each can come about as easily as any other" (*omnes casus aeque possibiles esse, seu pari facilitate evenire posse*). Laplace, on the other hand, in his *Essai philosophique*, states the following famous – and purely epistemic – criterion:

> The theory of chance consists in reducing all the events of the same kind to a certain number of cases equally possible, that is to say, to such as we may be equally undecided about in regard to their existence.... [Laplace 1952, p. 6]

> If [the various cases] are not [equally possible], we will determine first their respective possibilities, whose exact appreciation is one of the most delicate points of the theory of chance. [Laplace 1952, p. 11]

To assign equal probability to cases "such as we may be equally undecided about" is the notorious *principle of insufficient reason*. Although Laplace did not view it as controversial, many in the 19th century did. What determines when cases are equally probable, possible, or likely? This epistemological ambiguity in the meaning and determination of an FPS led inevitably to controversy in its application.

2. *D'Alembert and De Morgan.* For example, what is the chance of getting at least one head in two tosses of a fair coin? The standard solution to this problem regards the four possible outcomes of tossing a coin twice – HH, HT, TH, TT – as equally likely; since 3 out of these four cases are favorable, the probability is 3/4. In 1754, however, the French *philosophe* Jean Le Rond D'Alembert (1717–1783) advanced a different solution in his article 'Croix ou pile' in the *Encyclopedie*. D'Alembert reasoned that one would stop tossing the coin as soon as the desired head came up, so that there are really only three possible outcomes – H, TH, TT – two of which are favorable, and hence the probability is 2/3.

D'Alembert was far from being the first distinguished mathematician to make an elementary error of this type, but he is perhaps unique in the doggedness with which he subsequently defended his answer. Indeed, this was only the first of several instances where D'Alembert was led to disagree with the standard answers of the calculus of probabilities, and "with this article, the renowned mathematician opened a distinguished career of confusion over the theory of probabilities" (Baker 1975, p. 172).[16]

D'Alembert's criticisms were largely greeted with scorn and ridicule, but seldom seriously discussed. Laplace, for example, remarks that the probability would indeed be 2/3 "if we should consider with

D'Alembert these three cases as equally possible ..." (1952, p. 12), but he limits himself to giving the standard calculation without explaining why one set of equipossible cases is preferable to another.

The D'Alembert fallacy is possible because of the ambiguity in the concept of equipossibility and the Laplacean definition of probability. Laplace's treatment of these questions, although not confused, fails to come to grips with the fundamental issues. For one of the few serious discussions of D'Alembert's argument, one must turn to the writings of Augustus De Morgan, Laplace's most enthusiastic and influential English expositor during the first half of the 19th century.

De Morgan argued that there are essentially two very distinct considerations involved in the assessment of numerical probabilities. The first of these is *psychological*: the measurement and comparison of "the impressions made on our minds by different prospects," as in a judgment of equiprobability among alternatives. The second is *mathematical*: the rational use of such measures or comparisons, as in the computation of the probability of a complex event involving simpler, equiprobable outcomes. The two questions differ in that "any given answer to the first may admit of dispute," while "there is no fear of mathematics failing us in the second," (De Morgan 1845, p. 395).

Armed with this distinction, De Morgan was able to analyze the D'Alembert fallacy:

> [W]ith regard to the objection of D'Alembert ..., we must observe that if any individual really feel himself certain, in spite of authority and principle, as here laid down, that the preceding cases are equally probable, he is fully justified in adopting 2/3 instead of 3/4, till he see reason to the contrary, which it is hundreds to one he would find, if he continued playing for a stake throughout a whole morning, that is, accepting bets of two to one that H would not come up once in two throws, instead of requiring three to one.... The individual just supposed, has applied correct mathematics to a manner in which he feels obliged to view the subject, in which we think him wrong, but the error is in the first of the two considerations [above], and not in the second. [De Morgan 1845, p. 401]

Despite its interest, De Morgan's discussion is ultimately unsatisfactory. The choice of an FPS is described as a psychological consideration (which would suggest a subjectivist viewpoint), but the phrase "in which we think him wrong" suggests an objectivistic one. De Morgan appeals to experience to justify the classical choice of FPS in the D'Alembert problem, although probabilities for De Morgan were degrees of belief rather than empirical frequencies. The Laplacean view of probability was one of rational degree-of-belief, but his followers were understandably reluctant to uncouple probability from frequencies although, not surprisingly, unable to provide a logical description of the choice of FPS.

De Morgan later returned to the D'Alembert example in his *Formal Logic* (1847, pp. 199–200), and his brief discussion there is also interesting:

> [I]t may happen that the state of mind which *is*, is not the state of mind which should be. D'Alembert believed that it was *two* to *one* that the first head which the throw of a halfpenny was to give would occur before the third throw; a juster view of the mode of applying the theory would have taught him it was *three* to *one*. But he *believed* it, and thought he could show reason for his belief: to him the probability *was* two to one. But I shall say, for all that, that the probability *is* three to one: meaning, that in the universal opinion of those who examine the subject, the state of mind to which a person *ought* to be able to bring himself is to look three times as confidently upon the arrival as upon the non-arrival.

When De Morgan says that, for D'Alembert, "the probability *was*," the word probability is being used in a psychological or personalist sense; when he says "the probability *is*," the sense is logical or credibilist. But to say that the probability is three to one because that is "the universal opinion of those who examine the subject," while certainly candid, is hardly a devastating refutation of D'Alembert.

De Morgan deserves considerable credit for distinguishing between the psychological process of identifying a set of outcomes as equipossible, and the mathematical use of such a set to calculate probabilities, as well as his (implicit) distinction between the subjective and objective senses of probability. Where he fails is in his account of why the probability "is" three to one, and what empirical justification, if any, such a statement requires. These, however, were basic questions for which the theory of his day had no answer.

In the later half of the 19th century, a serious attack was mounted on epistemic probability and the principle of insufficient reason, and a direct confrontation with such questions could no longer be avoided.

6. THE PRINCIPLE OF INSUFFICIENT REASON

The contributions of Laplace represent a turning point in the history of probability. Before his work, the mathematical theory was (with the exception of the limit theorems of Bernoulli and DeMoivre) relatively unsophisticated, in effect a subbranch of combinatorics; its serious applications largely confined to games of chance and annuities. All this changed with Laplace. Not only did he vastly enrich the mathematical theory of the subject, both in the depth of its results and the range of the technical tools it employed, he demonstrated it to be a powerful instrument having a wide variety of applications in the physical and social sciences. Central to his system however, was the use of the so-called principle of insufficient reason.[17]

The 19th century debate about the validity of the principle of insufficient reason involved, of necessity, much broader issues. Is probability empirical or epistemic in nature? Can a probability be meaningfully assigned to any event? Are all probabilities numerically quantifiable? Beginning in the 1840's, and continuing on into the 20th century, a number of eminent British mathematicians, philosophers, and scientists began to address such questions, including De Morgan, Ellis, Mill, Forbes, Donkin, Boole, Venn, Jevons, MacColl, Edgeworth, Keynes, Ramsey, Jeffreys, and Broad.

1. *Donkin.* A comprehensive discussion of this literature would be beyond the scope of the present paper. Instead, we will confine our attention primarily to the contributions of William Fishburn Donkin, Savilian Professor of Astronomy in the University of Oxford from 1842 to 1869. Donkin wrote two papers on mathematical probability. One of these concerned the justification for the method of least squares and, although a valuable contribution to that subject, will not concern us here. The other paper is modestly titled 'On Certain Questions Relating to the Theory of Probabilities' (Donkin 1851). Donkin's paper, although little-known, is a lucid and careful attempt to clarify the foundations of the subject. It was written in response to criticisms by Forbes and others of Michell's argument that stars that are optically double are also physically so.

Donkin begins by stating that

> It will, I suppose, be generally admitted, and has often been more or less explicitly stated, that the subject matter of calculation in the mathematical theory of probabilities is *quantity of belief*.

There were some dissenters to this view of probability at the time Donkin wrote (e.g., Ellis 1844; Mill 1843), but they were few in number and, due at least in part to the influence of De Morgan, Laplace's views held sway in England.[18]

Donkin's philosophical view of probability may be summarized as *relative, logical, numerical, and consistent*. Probability is relative in the sense that it is never "inherent in the hypothesis to which it refers," but "always *relative* to a state of knowledge or ignorance." Nevertheless, Donkin was not a subjectivist, because he also believed probability to be

> *absolute* in the sense of not being relative to any individual mind; since, the same information being presupposed, all minds *ought* to distribute their belief in the same way.

Ultimately, any such theory of logical probability must resort to the principle of insufficient reason, and Donkin's was no exception. Indeed, if anything he saw its role as even more central to the theory than did Laplace:

> . . . the law which must always be made the foundation of the whole theory is the following: – *When several hypotheses are presented to our mind, which we believe to be mutually exclusive and exhaustive, but about which we know nothing further, we distribute our belief equally amonst them.*

Although Boole's detailed and influential criticism of the appeal to insufficient reason was still several years off (Boole 1854, pp. 363–375), Robert Leslie Ellis had already attacked its use on the grounds that it "erected belief upon ignorance" (Ellis 1850, p. 325). Donkin's response was to stake out a limited claim for the theory:

> [The force of] the argument commonly called the "sufficient reason" . . . in all cases depends (as it seems to me) upon a previous *assumption* that *an intelligible law exists* concerning the matter in question. If this assumption be admitted, and if it can be shown that there is only *one* intelligible law, then that must be the actual law. . . . A person who should dispute the propriety of dividing our belief equally amongst hypotheses about which we are equally ignorant, ought to be refuted by asking him to state *which is to be preferred*. He must either admit the proposed law, or maintain that there is no law at all.

This observation would not have disarmed Ellis, Boole, or Venn, who indeed denied the existence of any (determinate in the case of Boole) law at all. But it did draw the line clearly. Its vulnerability, as Boole realized, is simply that two or more sets of "mutually exclusive and exhaustive" hypotheses may present themselves "about which we know nothing further," and which give rise to incompatible probability assignments. Ramsey saw it as a virtue of the subjectivistic theory that it eluded this dilemma by dispensing with the requirement of a unique law, admitting more than one probability assignment as possible (Ramsey 1926, pp. 189–190).

But can one calculate probabilities no matter how complex the setting or information available? Cournot, for example, had earlier argued that there were three distinct categories of probability – objective, subjective, and philosophical, the last involving situations whose complexity precluded mathematical measurement.[19]

Donkin thought such arguments, essentially pragmatic in nature, not to the point:

> ... I do not see on what ground it can be doubted that every definite state of belief concerning a proposed hypothesis is in itself capable of being represented by a numerical expression, however difficult or impracticable it may be to ascertain its actual value.... [It is important to distinguish] the difficulty of *ascertaining numbers* in certain cases from a supposed difficulty of *expression by means of numbers*. The former difficulty is real, but merely relative to our knowledge and skill; the latter, if real, would be absolute, and inherent in the subject matter, which I conceive not to be the case.

This was an important distinction. It expresses a tenet of faith of logical probability: that all probabilities can, in principle be measured. On a basic philosophical level, such theories have never really answered Ramsey's simple criticism:

> It is true that about some particular cases there is agreement, but these somehow paradoxically are always immensely complicated; we all agree that the probability of a coin coming down heads is 1/2, but we can none of us say exactly what is the evidence which forms the other term for the probability relation about which we are then judging. If, on the other hand, we take the simplest possible pairs of propositions such as 'This is red', and 'That is blue', or 'This is red' and 'That is red', whose logical relations should surely be easiest to see, no one, I think, pretends to be sure what is the probability relation between them. [Ramsey 1926]

2. *Boole.* The first influential critic of the principle of insufficient reason was Boole. He says of its derivation:

> It has been said, that the principle involved in the above and in similar applications is that of the equal distribution of our knowledge, or rather of our ignorance – the assigning to different states of things of which we know nothing, and upon the very ground that we know nothing, equal degrees of probability. I apprehend, however, that this is an arbitrary method of procedure. [Boule 1854, p. 370]

As we have seen earlier (Section 3), to justify his criticism Boole pointed to instances where it was possible to partition the sample space of possible outcomes in different ways, each of which could plausibly be viewed as equipossible. Boole's criticisms, unfortunately, became more confusing as he attempted to clarify them. One might be forgiven, for example, for interpreting the passage just quoted as a clear rejection of the principle. But Boole later wrote:

> I take this opportunity of explaining a passage in the *Laws of Thought*, p. 370, relating to certain applications of the principle. Valid objection lies not against the principle itself, but against its application through arbitrary hypotheses, coupled with the assumption that any result thus obtained is necessarily the true one. The application of the principle employed in the text and founded upon the general theorem of development in Logic, I hold to be *not* arbitrary. [Boole 1862]

Perusal of "the application of the principle employed in the text" reveals it to be of the balls in an urn type, and what Boole now appears to be defending might be called the principle of *cogent* reason: if one possesses some information about the different alternatives, but this information is equally distributed amongst them, then one is justified in assigning the alternatives equal probability.

Boole appeared to regard both probabilistic independence (which he used extensively in his system) and uniformity of distribution as assumptions of *neutrality*, in each case a *via media* between conflicting extremes. There is a simple geometric sense in which this is true for the assumption of uniformity: the uniform distribution on $n + 1$ elements is the barycenter of the n-dimensional simplex of all probability distributions. But once more the consequences of a symmetry assumption lurk only partially visible. For depending on the use being made of a probability distribution, symmetrical or uniform distributions can often represent an extreme type of behavior. A good example of this involves the "birthday paradox": in a group of 23 or more people, the odds exceed 1/2 that at least two persons share a birthday in common (Feller 1968, p. 33). The calculation on which this statement is based assumes that births occur uniformly throughout the year. Although empirically false (see, e.g., Izenman and Zabell 1982), this does not affect the validity of the conclusion: the probability of a birthday "match" is minimized when the distribution of births is uniform (so that the probability of a match will be even greater under the true distribution).

It is difficult to assess Boole's immediate impact on his contemporaries. As the distinguished author of *The Laws of Thought*, his views on probability were certainly treated with respect. Nevertheless, they were highly idiosyncratic and confused in important respects.[20] Given the complexity and unattractiveness of his own system, and lacking the alternative philosophical foundation to the Laplacean edifice that was later provided by Venn's *Logic of Chance*, there was an obvious reluctance to abandon the classical theory. Nevertheless, his pointing to the fundamental ambiguity in the principle of insufficient reason was a lasting contribution, remembered long after the rest of his work on probability was forgotten.

Donkin represents what may be the highwater mark in the defense of the Laplacean position; Boole was its first influential English critic. After Boole and Venn the Laplaceans were on the defensive, first in the philosophical, later in the statistical and scientific communities. In response to the criticisms of Boole and his successors, many attempts were made to state unambiguous formulations of the principle of insufficient reason (e.g., by von Kries and Keynes), but their increasing obscurity and complexity ensured their rejection.[21]

The debate about the principle of insufficient reason and its consequence, Laplace's rule of succession, tapered off in the 1920s. This was partly because Ramsey's 1926 essay 'Truth and Probability' made the principle superfluous as a foundation for epistemic probability. When Fisher and Neyman produced statistical methodologies independent of the Bayes–Laplace edifice, Bayesian statistics essentially disappeared, only to be resuscitated by Savage nearly a quarter of a century later with the publication in 1954 of his *Foundations of Statistics*.

Savage's conversion to subjectivism occurred after he became acquainted with de Finetti's work, and his writings were largely responsible for bringing it into the mainstream of philosophical and statistical thought. At the center of de Finetti's system was the notion of exchangeability, and thus, initially exorcised, symmetry re-entered epistemic probability.

7. WHAT IS TO BE DONE?

Symmetry arguments are tools of great power; therein lies not only their utility and attraction, but also their potential treachery. When they are invoked one may find, as did the sorcerer's apprentice, that the results somewhat exceed one's expectations. Nevertheless, symmetry arguments enjoy an honored and permanent place in the arsenal of probability. They underlie the classical definition of probability that held sway for over two centuries, are central to virtually all quantitative theories of induction, appear as exchangeability assumptions in subjectivist theories, and, in the guise of group-invariance, still play an important role in modern theoretical statistics. Their use calls for judicious caution rather than benign neglect.

The ambiguity underlying the proper role of symmetry assumptions in the theory of probability stems in part from a corresponding ambiguity about the role the axioms play in the various axiomatic formulations of probability. Do the axioms enjoy a privileged status *vis-à-vis* their deducible consequences? Are they supposed to be intuitively more evident or simpler in form? If the justification for the axioms is their intuitive acceptability, what if some of their consequences violate those intuitions? As in so many cases, one can identify two polar positions on such issues, that of the *left-wing dadaists* and the *right-wing totalitarians*.[22]

The left-wing dadaists not only demand that the axioms be grounded in our intuitions, but that *all* deducible consequences of the axioms must be intuitively acceptable as well. Intuitive acceptability was the warrant for the axioms in the first place, and since there is no obvious reason to favor certain intuitions over others, all must be satisfied. If the consequences of a set of axioms violate our intuitions, then those axioms must be abandoned and replaced. A leading exponent of this position is L. Jonathan Cohen.[23]

The problem with such a position is that our intuitions, or at least our *untutored* intuitions, are often mutually inconsistent and any consistent theory will necessarily have to contradict some of them. During the last two decades many psychologists, notably Daniel Kahneman and Amos Tversky, have demonstrated that popular intuitions are often inconsistent not merely with the standard axioms of probability, but with essentially *any* possible axiomatization of probability; that "people systematically violate principles of rational decision-making when judging probabilities, making predictions, or otherwise attempting to cope with probabilistic tasks" (Slovic, Fischhoff, and Lichtenstein 1976).[24]

The right-wing totalitarians, on the other hand, believe that once an axiom system is adopted, one must accept without question every consequence that flows from it. One searches within one's heart, discovers the basic properties of belief and inference, christens them axioms, and then all else follows as logical consequence. Once the axioms are brought to the attention of unbelievers, they must, like Saul on the road to Damascus, be smitten by instantaneous conversion or they stand convicted of irrational obtuseness. One representative of this position is E. T. Jaynes, who dates his adherence to Bayesianism to the time when he encountered Cox's axiomatization of epistemic probability, and who views the Shannon axioms for entropy as an unanswerable foundation for his method of maximum entropy.[25]

This position errs in giving the axioms too distinguished a position, just as the previous position gave them too little. A set of axioms A, together with $T(A)$, the theorems deducible from it, forms a self-consistent whole S. Let us say that any subset $B \subseteq S$, such that $B \cup T(B) = S$, is an *axiom-system for S*. Mathematically speaking, all possible axiom-systems for S must be regarded as starting out on an equal footing, and which axiom-system is ultimately chosen is essentially a matter of preference, depending on considerations such as simplicity, elegance, and intuitive acceptability.

The key point is that having tentatively adopted an axiom system, one is not obligated to uncritically accept its consequences. In both formal logic and subjective probability, the theory polices sets of beliefs by testing them for inconsistencies, but it does not dictate how detected inconsistencies should be removed. If, as was the case with some of the symmetry assumptions previously discussed, the consequences are deemed unacceptable, then the assumption will be discarded. If, on the other hand, the axioms seem compelling, as in mathematical probability, then surprising consequences such as the birthday paradox will be regarded as valuable correctives to our erroneous, untutored intuitions; that is why the theory is useful. What is or should be at play is a dynamic balance. As Nelson Goodman argues:

> Inferences are justified by their conformity to valid general rules, and . . . general rules are justified by their conformity to valid inferences. But this circle is a virtuous one. The point is that rules and particular inferences alike are justified by being brought into agreement with each other. *A rule is amended if it yields an inference we are unwilling to accept; an inference is rejected if it violates a rule we are unwilling to amend*. The process of justification is the delicate one of making mutual adjustments between rules and accepted inferences; and in the agreement achieved lies the only justification needed for either [Goodman 1979, p. 64].

Symmetry assumptions must therefore be tested in terms of the particular inferences they give rise to. But – and this is the rub – particular inferences can only be reasonably judged in terms of particular situations, whereas symmetry assumptions are often proposed in abstract and theoretical settings devoid of concrete specifics.[26]

Fundamentally at issue here are two very different approaches to the formulation of a logic of probability. Extreme subjectivists adopt a *laissez faire* approach to probability assignments, emphasizing the unique aspects attending the case at hand. They do not deny the utility of symmetry arguments, but, as Savage remarks, they "typically do not find the contexts in which such agreement obtains sufficiently definable to admit of expression in a postulate" (Savage 1954, p. 66). Such arguments fall instead under the rubric of what I. J. Good terms "suggestions for using the theory, these suggestions belonging to the technique rather than the theory" itself (Good 1952, p. 107).

Proponents of logical theories, in contrast, believe (at least in principle) that if the evidence at one's disposal is stated with sufficient precision in a sufficiently rich language then agreement can be forced via considerations of symmetry. At the level of ordinary language such claims founder at the very outset on Ramsey's simple objection (quoted earlier in Section 6). Instead, simple model languages are introduced and probabilities computed "given" statements descriptive of our state of knowledge. Such formal systems do not escape subjectivism, they enshrine it in the equiprobable partitions assumed.

Practical attempts to apply logical probability always seem to lead back to discussions about events "concerning the probability of which we absolutely know nothing antecedently to any trials made concerning it." Such attempts are ultimately divorced from reality, if only because understanding the very meaning of the words employed in describing an event already implies knowledge about it. Thus, it is not surprising that the three leading 20th century proponents of logical probability – Keynes, Jeffreys, and Carnap – all eventually recanted to some extent or another.[27] Carnap, for example, wrote

> I think there need not be a controversy between the objectivist point of view and the subjectivist or personalist point of view. Both have a legitimate place in the context of our work, that is, the construction of a system of rules for determining probability values with respect to possible evidence. At each step in the construction, a choice is to be made; the choice is not completely free but is restricted by certain boundaries. Basically, there is merely a difference in attitude or emphasis between the subjectivist tendency to emphasize the existing freedom of choice, and the objectivist tendency to stress the existence of limitations. [Carnap 1980, p. 119]

This little-known, postumously published passage is a substantial retreat from the hard-core credibilism of the *Logical Foundations of Probability*. But it was inevitable. Symmetry arguments lie at the heart of probability. But they are tools, not axioms, always to be applied with care to specific instances rather than general propositions.

8. ENVOI

As a final illustration of the seductive nature of symmetry arguments in probability, and as a challenge to the reader, I end with a little puzzle, which I will call the *exchange paradox*:[28]

> *A*, *B*, and *C* play the following game. *C* acts as referee and places an unspecified amount of money *x* in one envelope and amount 2*x* in another envelope. One of the two envelopes is then handed to *A*, the other to *B*.
>
> *A* opens his envelope and sees that there is \$10 in it. He then reasons as follows: "There is a 50–50 chance that *B*'s envelope contains the lesser amount *x* (which would therefore be \$5), and a 50–50 chance that *B*'s envelope contains the greater amount 2*x* (which would therefore be \$20). If I exchange envelopes, my expected holdings will be (1/2)\$5 + (1/2)\$20 = \$12.50, \$2.50 in excess of my present holdings. Therefore I should try to exchange envelopes."
>
> When *A* offers to exchange envelopes, *B* readily agrees, since *B* has already reasoned in similar fashion.

It seems unreasonable that the exchange be favorable to both, yet it appears hard to fault the logic of either. I will resist the temptation to explain what I take to be the resolution of the paradox, other than noting that all hinges on *A*'s apparently harmless symmetry assumption that it is equally likely that *B* holds the envelope with the greater or the lesser amount.[29]

Department of Mathematics
Northwestern University

NOTES

[1] The symbol $\binom{n}{k}$ denotes the binomial coefficient $n!/[k!(n-k)!]$. Note that in the theorem the sequence is assumed to be infinite; this requirement is sometimes overlooked, although it is necessary for the general validity of the theorem.

[2] There also exist finite forms of de Finetti's theorem, which permit one to dispense with the assumption that the number of trials is infinite. In such cases the integral mixture is either replaced by a discrete sum or serves as an approximation to the exact probability; see Diaconis and Freedman (1980a).

[3] The existence of limiting frequencies for infinite exchangeable sequences follows from their stationarity, and is an immediate consequence of the ergodic theorem; see, e.g., Breiman (1968, p. 118, Theorem 6.28).

[4] For further discussion of Bayes's scholium, see Murray (1930), Edwards (1978). For an interesting account of how Bayes's argument has often been misconstrued by statisticians to fit their foundational preconceptions, see Stigler (1982).

[5] It is an interesting historical footnote that Johnson's derivation almost never appeared. After the appearance of the third volume of his *Logic* in 1924, Johnson began work on a fourth volume, to be devoted to probability. Unfortunately, Johnson suffered a stroke in 1927, and the projected work was never finished. Drafts of the first three chapters were edited by R. B. Braithwaite and published posthumously as three separate papers in *Mind* during 1932. Johnson's mathematical derivation of the continuum of inductive methods from the sufficientness postulate appeared as an appendix in the last of the three. G. E. Moore, then editor of *Mind*, questioned whether so technical a result would be of general interest to its readership, and it was only on the insistence of Braithwaite that the appendix was published (Braithwaite 1982, personal communication).

[6] For further information about Johnson's sufficientness postulate, and a complete version of his proof, see Zabell (1982).

[7] In brief, this is because even when one lacks information about specific, identifiable categories, one may possess information about the vector of *ordered* probabilities. (For example, one may know that a die is biased in favor of one face, but not know which face it is.)

[8] See generally Heath (1913, Chapter 4); Kahn (1960); Dicks (1970, Chapter 3). For the original Greek texts of the fragments of Anaximander, with accompanying English translation, commentary, and discussion, see Kirk and Raven (1957, Chapter 3).

[9] Perhaps the most pessimistic assessment of the state of our information is that of Dicks (1970, pp. 45–46).

[10] In its general form (neither of two exactly symmetrical alternatives will occur), it also crops up from time to time in 19th century philosophical discussions of probability. Two examples are (1) Bolzano: "... if we are to have a rational expectation that a certain result will take place, for example that Caius will draw a certain ball from several balls in an urn, then we must presuppose that the relation between these balls and Caius is such that the reasons for drawing that particular ball are not exactly like the reasons for drawing some other ball, since otherwise he wouldn't draw any" (Bolzano 1837, p. 245 of 1972 edition.); (2) Cook Wilson: "... if a number of cases, mutually exclusive ..., were in the nature of things equally possible, not one of them could happen. If the claim of any one of them in reality were satisfied, so must the claim of any other, since these claims are equal, and therefore if one happens all must, but by hypothesis if one happens no other can; thus the only possible alternative is that none of them can happen" (Wilson 1900, p. 155).

[11] Aristocles, quoted in Long (1974, p. 81); c.f. Diogenes Laertius, *Life of Pyrrho* 9.107; Sextus Empiricus, *Outlines of Pyrrhonism* 1.8. For general information on the Pyrrhonian skeptics, see Stough (1969, Chapter 2); Long (1974, pp. 75–88). The *ou mallon* argument itself is discussed in some detail by DeLacy (1958).

[12] See generally Stough (1969, pp. 50–66); Long (1974, pp. 95–99).

[13] See, e.g., Plato, *Laws* 709, 889 b–d; Aristotle, *Metaphysics* 1070ab. (Strictly speaking, Aristotle distinguishes between *automaton* (chance, spontaneity) and *tyche* (luck, fortune).

[14] For further discussion of Arbuthnot, see Hacking (1965, pp. 75–77); Hacking (1975, Chapter 18); Pearson (1978, pp. 127–133, 161–162).

[15] For a recent and very readable account of the dispute, see Gower (1982). Similar issues arose in later discussions of geometrical probability: what does it mean to select points (or lines, or triangles) at random? Venn (1888, pp. 100–101), reporting one such discussion, quotes the English mathematician Crofton as asserting that "at random" has "a very clear and definite meaning; one which cannot be better conveyed than by Mr Wilson's definition, 'according to no law'. . . ." "Mr. Crofton holds," Venn continues, "that any kind of *unequal* distribution [of points in a plane] would imply law," to which Venn retorts, "Surely if they tend to become *equally* dense this is just as much a case of regularity or law." Where James Bernoulli had attempted to subsume the probability of causes under that of chances (to use Hume's terminology), the frequentist Venn subsumes the probability of chances under that of causes.

[16] See generally Todhunter (1865, Chapter 13); Baker (1975, pp. 171–180); Pearson (1978, Chapter 12). For a recent but unconvincing attempt at rehabilitation, see Daston (1979).

[17] Laplace nowhere actually uses this term, which is of later origin. Writing in 1862, Boole refers to "that principle, more easily conceived than explained, which has been differently expressed as the 'principle of non-sufficient reason', the principle of equal distribution of knowledge or ignorance' [footnote omitted], and the 'principle of order'," (Boole 1862).

[18] When Donkin wrote his paper the first frequentist theories (apart from occasional allusions in the earlier literature) were less than a decade old. As Porter (1986, p. 77) notes, "in 1842 and 1843, four writers from three countries independently proposed interpretations of probability that were fundamentally frequentist in character." These four – Jakob Friedrick Fries in Germany, Antoine Augustin Cournot in France, and Richard Leslie Ellis and John Stuart Mill in England – were the harbingers of an increasingly empirical approach to probability. (Curiously, after correspondence with the astronomer John Herschel, Mill actually withdrew his objections to Laplace's epistemic view of probability from the second (1846) and later editions of his *Logic*; see Strong (1978).) Despite this early efflorescence, the frequency theory did not begin to gain widespread acceptance until its careful elaboration, nearly a quarter of a century later, in John Venn's *Logic of Chance* (1st ed. 1866). For discussion of the work of Fries, Cournot, Ellis, and Mill, see Porter (1986, pp. 77–88), Stigler (1986, pp. 195–200); for discussion of Venn's *Logic*, Salmon (1980).

[19] The argument that some probabilities are "philosophical" (i.e., inherently non-numerical) was often made by those who thought the mathematical theory had outreached its grasp. Strong (1976, p. 207, n. 5) notes the use of the distinction in K. H. Frömmichen's 1773 work, *Über die Lehre des Wahrscheinlich*, "the earliest . . . that I have been able definitely to date," as well the better known treatment in Kant's *Logik* of 1781. See von Wright (1957, p. 217, n. 9) for further references to the 19th century

literature. In addition to the names given there, one could add those of the Scottish philosopher Dugald Stewart and the English jurists Starkie, Wills, and Best. For the related criticisms of the French positivists Destutt de Tracy, Poinsot, and Comte, see Porter (1986, p. 155) and Stigler (1986, pp. 194–195).

[20] Many of these are touched on by Keynes in scattered passages throughout his *Treatise on Probability* (1921). Hailperin (1976) is a useful attempt at rational reconstruction. For discussion of Boole's criticism of the Laplace/De Morgan analysis of inductive reasoning in terms of probability, see the excellent article of Strong (1976).

[21] See generally Keynes (1921, Chapters 4 and 6).

[22] There is obviously an element of intentional caricature in what follows, although perhaps less than might be supposed.

[23] "... ordinary human reasoning ... cannot be held to be faultily programmed: it sets its own standards" (Cohen 1981, p. 317).

[24] Much of this work is summarized in Kahneman, Slovic, and Tversky (1982).

[25] Although not readily available, Jaynes's early Socony Mobil Oil lecture notes (Jaynes 1958) provide a vigorous and very readable exposition of his viewpoint.

[26] There are some notable exceptions to this. W. E. Johnson, for example, in discussing his sufficientness postulate, argued that:

"the postulate adopted in a controversial kind of theorem cannot be generalized to cover all sorts of working problems; so it is the logician's business, having once formulated a specific postulate, to indicate very carefully the factual and epistemic conditions under which it has practical value." (Johnson 1932, pp. 418–419)

[27] For Keynes's recantation, see Good (1965, p. 7). In the third edition of his book *Scientific Inference*, Jeffreys suggests that in controversial cases the appropriate choice of reference prior could be decided by an international panel of experts. Such a position is obviously incompatible with credibilism as usually understood. For Carnap, see the text *infra*.

[28] I first heard the paradox from Steve Budrys of the Odesta Corporation, on an otherwise unmemorable night at the now defunct Chessmates in Evanston. It does not originate with him, but I have been unable to trace its ultimate source.

Note added in proof: Persi Diaconis and Martin Gardner inform me that the paradox is apparently due to the French mathematician Maurice Kraitchik; see Maurice Kraitchik, *Mathematical Recreations*, 2nd ed. (New York: Dover, 1953), pp. 133–134. In Kraitchik's version two persons compare their neckties, the person with the less valuable necktie to receive both.

[29] I thank Persi Diacanis, David Malament and Brian Skyrms for helpful comments.

BIBLIOGRAPHY

Arbuthnot, John (1711) 'An argument for divine providence taken from the constant regularity observed in the births of both sexes', *Philosophical Transactions of the Royal Society of London* **27**, 186–190.

Baker, Keith Michael (1975) *Condorcet: From Natural Philosophy to Social Mathematics* (Chicago: University of Chicago Press).

Bayes, Thomas (1764) 'An essay towards solving a problem in the doctrine of chances', *Philosophical Transactions of the Royal Society of London* **53**, 370–418.

Bolzano, Bernard (1837) *Wissenschaftslehre*. Translated 1972 under the title *Theory of Science* (R. George, ed. and trans.) (Berkeley and Los Angeles: University of California Press).

Boole, George (1854) *An Investigation of the Laws of Thought* (London: Macmillan.) (Reprinted 1958, New York: Dover Publications.)

Boole, George (1862) 'On the theory of probabilities', *Philosophical Transactions of the Royal Society of London* **152**, 386–424.

Breiman, Leo (1968) *Probability* (Reading, Mass.: Addison-Wesley).

Broad, C. D. (1975) *Leibniz: An Introduction* (Cambridge University Press).

Carnap, Rudolph (1950) *Logical Foundations of Probability* (The University of Chicago Press. Second edition, 1960).

Carnap, Rudolph (1952) *The Continuum of Inductive Methods* (Chicago: University of Chicago Press).

Carnap, Rudolph (1980) 'A basic system of inductive logic, part II', in *Studies in Inductive Logic and Probability*, volume II (Richard C. Jeffrey, ed.) (Berkeley and Los Angeles: University of California Press) pp. 7–155.

Cohen, L. Jonathan (1981) 'Can human irrationality be experimentally demonstrated?', *The Behavioral and Brain Sciences* **4**, 317–370 (with discussion).

Cournot, Antoine Augustin (1843) *Exposition de la théorie des chances et des probabilités* (Paris: Libraire de L. Hachette).

Daston, Lorraine J. (1979) 'D'Alembert's critique of probability theory', *Historia Mathematica* **6**, 259–279.

De Finetti, Bruno (1937) 'La prévision: ses lois logiques, ses sources subjectives', *Annales de l'Institut Henri Poincaré* **7**, 1–68.

DeLacy, Phillip (1958) 'Ou mallon and the antecedents of ancient scepticism', *Phronesis* **3**, 59–71.

De Moivre, Abraham (1756) *The Doctrine of Chances* (3rd ed.), London.

De Morgan, Augustus (1845) 'Theory of probabilities', *Encyclopedia Metropolitana, Vol. 2: Pure Mathematics* (London: B. Fellowes *et al.*) pp. 393–490.

De Morgan, Augustus (1847) *Formal Logic: Or the Calculus of Inference Necessary and Probable*. London.

Diaconis, Persi (1977) 'Finite forms of de Finetti's theorem on exchangeability', *Synthese* **36**, 271–281.

Diaconis, Persi and Freedman, David (1980a) 'Finite exchangeable sequences', *Annals of Probability* **8**, 745–764.

Diaconis, Persi and Freedman, David (1980b) 'De Finetti's theorem for Markov chains', *Annals of Probability* **8**, 115–130.

Diaconis, Persi and Freedman, David (1980c) 'De Finetti's generalizations of exchangeability', in *Studies in Inductive Logic and Probability*, volume II (Richard C. Jeffrey, ed.) (Berkeley and Los Angeles: University of California Press) pp. 233–249.

Diaconis, Persi and Freedman, David (1985) 'Partial exchangeability and sufficiency', *Statistics: Applications and New Directions*. In *Proceedings of the Indian Statistical Institute Golden Jubilee International Conference* (Calcutta: Indian Statistical Institute) pp. 205–236.

Dicks, D. R. (1970) *Early Greek Astronomy to Aristotle* (Ithaca: Cornell University Press).

Donkin, William Fishburn (1851) 'On certain questions relating to the theory of probabilities', *Philosophical Magazine* **1**, 353–368; **2**, 55–60.

Edwards, A. W. F. (1978) 'Commentary on the arguments of Thomas Bayes', *Scandinavian Journal of Statistics* **5**, 116–118.

Ellis, Richard Leslie (1844) 'On the foundations of the theory of probabilities', *Transactions of the Cambridge Philosophical Society* **8**, 1–6.

Ellis, Richard Leslie (1850) 'Remarks on an alleged proof of the "method of least squares" contained in a late number of the *Edinburgh Review*', *Philosophical Magazine* **37**, 321–328.

Feller, William (1968) *An Introduction to Probability Theory and Its Applications*, vol. 1, 3rd ed. (New York: Wiley).

Good, Irving John (1952) 'Rational decisions', *Journal of the Royal Statistical Society B* **14**, 107–114.

Good, Irving John (1965) *The Estimation of Probabilities: An Essay on Modern Bayesian Methods* (Cambridge, Mass.: M. I. T. Press).

Goodman, Nelson (1979) *Fact, Fiction, and Forecast* (3rd ed.) (Indianapolis: Hackett Publishing Company).

Gower, Barry (1982) 'Astronomy and probability: Forbes versus Michell on the distribution of the stars', *Annals of Science* **39**, 145–160.

Hacking, Ian (1965) *Logic of Statistical Inference* (Cambridge University Press).

Hacking, Ian (1967) 'Slightly more realistic personal probability', *Philosophy of Science* **34**, 311–325.

Hacking, Ian (1975) *The Emergence of Probability* (Cambridge University Press).

Hailperin, Theodore (1976) *Boole's Logic and Probability*. Studies in Logic and the Foundations of Mathematics, volume 85 (Amsterdam: North-Holland).

Heath, Sir Thomas (1913) *Aristarchus of Samos: The Ancient Copernicus* (Oxford: The Clarendon Press). (Reprinted 1981, New York: Dover Publications.)

Hewitt, Edwin and Savage, Leonard J. (1955) 'Symmetric measures on Cartesian products', *Transactions of the American Mathematical Society* **80**, 470–501.

Hume, David (1739) *A Treatise of Human Nature*. London. (Page references are to the 2nd edition of the L. A. Selbe-Bigge text, revised by P. H. Nidditch, Oxford: The Clarendon Press, 1978.)

Hussey, Edward (1972) *The Presocratics* (New York: Charles Scribner's Sons).

Izenman, Alan J. and Zabell, Sandy L. (1981) 'Babies and the blackout: The genesis of a misconception', *Social Science Research* **10**, 282–299.

Jaynes, Edwin T. (1958) *Probability Theory in Science and Engineering*. Colloquium Lectures in Pure and Applied Science, no. 4 (Dallas: Socony Mobil Oil).

Jeffrey, Richard C. (1977) 'Mises redux', *Basic Problems in Methodology and Linguistics: Proceedings of the Fifth International Congress of Logic, Methodology and Philosophy of Science, Part III* (R. Butts and J. Hintikka, eds.) (Dordrecht: D. Reidel).

Johnson, William Ernest (1924) *Logic, Part III: The Logical Foundations of Science* (Cambridge University Press).

Johnson, William Ernest (1932) 'Probability: The deductive and inductive problems', *Mind* **49**, 409–423.

Kahn, Charles H. (1960) *Anaximander and the Origins of Greek Cosmology* (New York: Columbia University Press).

Kahneman, D., Slovic, P., and Tversky, A. (1982) *Judgment Under Uncertainty: Heuristics and Biases* (Cambridge University Press).

Keynes, John Maynard (1921) *A Treatise on Probability* (London: Macmillan).

Kirk, G. S. and Raven, J. E. (1957) *The Presocratic Philosophers: A Critical History with a Selection of Texts* (Cambridge University Press).

Kyburg, Henry (1974) 'Propensities and probabilities', *British Journal for the Philosophy of Science* **25**, 358–375.

Laplace, Pierre Simon Marquis de (1952) *A Philosophical Essay on Probabilities* (F. W. Truscott and F. L. Emory, trans.) (New York: Dover Publications).

Long, A. A. (1974) *Hellenistic Philosophy: Stoics, Epicureans, Sceptics* (New York: Charles Scribner's Sons).

Mill, John Stuart (1843) *A System of Logic*, 2 vols. London.

Michell, J. (1767) 'An inquiry into the probable parallax and magnitude of the fixed stars from the quantity of light which they afford to us, and the particular circumstances of their situation', *Philosophical Transactions of the Royal Society* **57**, 234–264.

Murray, F. H. (1930) 'Note on a scholium of Bayes', *Bulletin of the American Mathematical Society* **36**, 129–132.

Oresme, Nicole (1966) *De proportionnibus proportionum and Ad pauca respicientes* (E. Grant ed. and trans.) (University of Wisconsin Press).

Owen, G. E. L. (1966) 'Plato and Parmenides on the timeless present', *The Monist* **50**, 317–340.

Pearson, Karl (1978) *The History of Statistics in the 17th and 18th Centuries* (E. S. Pearson, ed.) (New York: Macmillan).

Porter, Theodore (1986) *The Rise of Statistical Thinking* (Princeton University Press).

Ramsey, Frank Plumpton (1926) 'Truth and probability', in *The Foundations of Mathematics and Other Logical Essays* (R. B. Braithwaite, ed.) (London: Routledge and Kegan Paul, 1931) pp. 156–198.

Salmon, Wesley C. (1980) 'John Venn's *Logic of Chance*', in *Pisa Conference Proceedings*, vol. 2 (J. Hintikka, D. Gruender, and E. Agazzi, eds.) (Dordrecht: D. Reidel).

Savage, Leonard J. (1954) *The Foundations of Statistics* (New York: John Wiley) (Reprinted 1972, New York: Dover).

Slovic, P., Fischhoff, B., and Lichtenstein, S. (1976) 'Cognitive processes and societal risk taking', in J. S. Carroll and J. W. Payne (eds.), *Cognition and Social Behavior* (Hillsdale, N.J.: Erlbaum).

Stigler, Stephen M. (1982) 'Thomas Bayes's Bayesian inference', *Journal of the Royal Statistical Society* Series A **145**, 250–258.

Stigler, Stephen M. (1986) *The History of Statistics* (Harvard University Press).

Stough, Charlotte L. (1969) *Greek Skepticism* (University of California Press).

Strong, John V. (1976) 'The infinite ballot box of nature: De Morgan, Boole, and Jevons on probability and the logic of induction', *PSA 1976: Proceedings of the Philosophy of Science Association* **1**, 197–211.

Strong, John V. (1978) 'John Stuart Mill, John Herschel, and the "probability of causes"', *PSA 1978: Proceedings of the Philosophy of Science Association*, **1**, 31–41.

Todhunter, Isaac (1865) *A History of the Mathematical Theory of Probability from the Time of Pascal to That of Laplace* (London: Macmillan) (Reprinted 1965, New York: Chelsea.)

Venn, John (1858) *The Logic of Chance* (3rd ed.) (London: Macmillan).

Wilson, John Cook (1900) 'Inverse or "a posteriori" probability', *Nature* **63**, 154–156.

von Wright, Georg Henrik (1957) *The Logical Problem of Induction* (2nd revised edition). (New York: Macmillan.)

Zabell, Sandy L. (1982) 'W. E. Johnson's "sufficientness" postulate', *Annals of Statistics* **10**, 1091–1099.

HAIM GAIFMAN

A THEORY OF HIGHER ORDER PROBABILITIES[1]

INTRODUCTION

The assignment of probabilities is the most established way of measuring uncertainties on a quantitative scale. In the framework of subjective probability, the probabilities are interpreted as someone's (the agent's) degrees of belief. Since justified belief amounts to knowledge, the assignment of probabilities, in as much as it can be justified, expresses knowledge. Indeed, knowledge of probabilities appears to be the basic kind of knowledge that is provided by the experimental sciences today.

This is knowledge of a partial, or incomplete, nature, but not in the usual sense of "partial". Usually we mean by "partial knowledge" knowledge of some, but not all, of the facts. But knowing that a given coin is unbiased does not enable one to deduce any non-tautological fact concerning the results of the next, say fifty tosses; every sequence of outcomes is possible. And yet it constitutes very valuable knowledge about these very same outcomes.

What is the objective content of this knowledge? What kind of fact is the fact that the true probability of "heads" is 0.5, i.e., that the coin is unbiased? I have argued elsewhere, (1983), that rather than to classify subjective and objective probabilities as two different kinds we should do better to regard them as two extremes of a spectrum. In that paper I considered the following question: Assuming a probability distribution which represents someone's beliefs, what is it that makes this distribution "objective"? As a way of answering it I pointed out and analyzed two aspects of objectiveness: inner stability and success. To go into these points here would make for too long a digression. So I shall start by taking it for granted that certain probability assignments are regarded by us as expressing fuller knowledge than other assignments. We also think that these "better" assignments are more likely to succeed, or to be in tune with the actual world. I shall not elaborate here on what constitutes "being in tune with the world". Let me only point out that the notion can be given precise meaning, as is illustrated

Brian Skyrms and William L. Harper (eds.), Causation, Chance, and Credence, Vol. I, 191–219.

by the concept of a calibrated prior, cf. (Dawid 1982). We do require, for example, of a successful weather forecaster that, if he has predicted on many occasions "30% rain", then among these occasions the number of cases in which it *did* rain afterwards is sufficiently near 30%.

Thus, we subject our probability assignments to judgment and evaluation and we are ready to change them accordingly. Having, for example, to estimate the possibility of a rain tomorrow, I might give it, going by today's weather, 70%. But I shall be highly uncertain about my assignment and will adopt the different value given, five minutes later, in the weather forecast.

When the uncertainty of probabilistic judgments is measured by assigning probabilities to them we get second order probabilities. An example of a second order probability is furnished by a cartoon in *The New Yorker* showing a forecaster making the following announcement:

There is now 60% chance of rain tomorrow, but, there is 70% chance that later this evening the chance of rain tomorrow will be 80%.

Formally, the process of assigning probabilities to probabilistic assignments can be iterated any number of times, just as a modality operant can. The goal of this paper is to present a general and adequate semantics for higher order probabilities and to obtain, via representation theorems, nice easily understood structures which give us a handle on the situation.

To give concrete meaning to our system, imagine that the agent who assigns probabilities to events has in mind also another assignment which derives from a more comprehensive state of knowledge. (When I say "more comprehensive state of knowledge" I do not only mean that some additional specific facts are known; I include also knowledge resulting from greater experience, and I leave aside the question whether having the experience can be represented as knowing some facts.) This second assignment, currently unknown to him, can be construed in various ways. If the agent believes in objective probabilities he may take the second assignment to be the objective probability distribution. But such a strong assumption is not necessary. The second assignment can be that of an expert, whose verdict the agent is willing to accept. Even simpler – it can be the agent's own assignment at some future time, say next morning. Thus the future time agent can play the role of the expert with respect to the present time one. Common to all these conceptions is a certain operational meaning that the second assignment has for the agent's behavior. Roughly put, it is: *The mere knowledge of the second assignment will make the agent adopt it as his subjective probability*. This general indication should for the moment suffice. Later, the operational meaning will be made precise when we use it in a coherence argument to derive a crucial axiom.

Let F be the field of events assigned probabilities by the agent. Then, with every event, A, in F and every real closed interval Δ we correlate another event, $pr(A, \Delta)$, *in the same field* F: the event that, in the second assignment, A's probability lies in Δ.

If P is the agent's probability function then the forecast in *The New Yorker* cartoon can be summed up as follows, where A = 'tomorrow it will rain', the agent is the forecaster at his current state, and pr represents his state later in the evening:

$$P(A) = 0.6 \qquad P(pr(A, [0.8, 0.8])) = 0.7$$

F, $P(\ )$ and $pr(\ ,\ )$ form what I shall call a *simple HOP* (Higher Order Probability structure).

The formalism of HOPs can accommodate various interpretations. The terms 'expert assignment' or 'expert probability' shall be used as convenient names for the second assignment. We do not presuppose any particular view concerning the nature of the "expert". As pointed out, the relevance of the expert probability is its operational meaning for the agent's behavior. *It is up to us (or to the agent) to decide what in the given context constitutes an "expert"*. If "expert knowledge" includes all the facts then, of course, the expert probabilities can be only 0 or 1 and A occurs iff its probability is 1; hence, in this case $pr(A, [1, 1]) = A$ and $pr(A, [0, 0]) = -A$. At the other extreme, the agent regards himself as being the expert, i.e., the second assignment is P; this leads to the "opposite" trivialization of the HOP, where for each A we have: $P(pr(A, \Delta)) = 1$ if $P(A) \in \Delta$, $P(pr(A, \Delta)) = 0$ – otherwise. Generally, the agent will regard the expert as being more knowledgeable than himself, but not omniscient.

We shall show that a certain well-motivated coherence principle is equivalent to constraining the HOP so that the agent's probability P is a mixture of mutually orthogonal expert probabilities. Thus, the well-known "dualistic" view whereby a subjective probability is a subjective mixture of unknown objective probabilities is derivable from a general coherence principle.

In its general form a HOP contains also a partially ordered set of *stages* and *pr* is a 3-place operator: $pr(A, s, \Delta)$ *is the event that the probability of A at stage s lies in* Δ. The partial order reflects knowledge inclusions: $s \leqslant t$ if the knowledge at stage t includes the knowledge at stage s. If it is total, the stages can be regarded as temporal phases of a single expert whose knowledge is non-decreasing with time. But in general the order is only partial. It can also be empty. We can thus represent many "incomparable" experts which, in the context of knowledge in distributed systems, is the many-player, or multi-processor case.

We shall first establish the basic properties of simple HOPs. Then, going on to general HOPs, we shall derive the basic properties concerning stages that are comparable under the partial ordering. Additional notions and principles which concern incomparable stages will have to be considered in further developments of the framework; this is a natural line of future research.

We shall also define, in a separate section, a formal logical calculus, to be called *probability logic*, which is naturally associated with HOPs. Various modalities can be reconstructed within this calculus. In particular, S5 is naturally embeddable in the probability logic of simple HOPs; the many-player generalization of S5, $S5_{(m)}$, considered first by Hintikka (1962) (cf. Halpern-Moses (1985)) is naturally embeddable in the probability logic for general HOPs where the partial ordering of the stages is empty.

The import of the subject for various branches of philosophy and for the foundations of probability is obvious. Also obvious should be its bearing upon applied probabilistic reasoning in distributed networks, or upon effects to incorporate such reasoning in AI systems. Our goal has been first and foremost to clarify some basic concepts and to outline a general, conceptually clean, framework. In such a framework one can use freely and to good effect statements such as: 'Alice knows with certainty at least 0.7 Bob's probability for the event X, within error $\leqslant 0.05$'. Here the "stages" are Alice and Bob. Writing the stage argument as an index, this is expressed in our formalism as the claim that for every x

$$pr_{\text{Bob}}(X, [x, x]) \subset pr_{\text{Alice}}(pr_{\text{Bob}}(X, [x - 0.05, x + 0.05]), [0.7, 1])$$

(I.e., if Bob's probability for X is x then Alice assigns probability at least 0.7 to the event that Bob's probability for X lies in $[x - 0.05, x +$

0.05]; the "if ... then ..." is expressed as an inclusion between events.) Such statements express intuitive thinking which may underlie involved technical proofs; to use them openly and precisely can help us as a guide for finding and organizing our arguments.

A theoretic framework for higher order probabilities yields also insights into systems of reasoning which extend Bayesian methods. For example, Jeffrey's probability kinematics, (1965), is a way of carrying over changes in the probability, from some finite field of events F, to the field generated by F and some other field F^*. In terms of a HOP, a change in the probabilities over F is achieved by conditioning on Boolean combinations of events of the form $pr(\mathrm{A}, \Delta)$, where A ranges over F. A necessary and sufficient condition for obtaining Jeffrey's kinematics with respect to every change over F is that for every event A which is atomic in F, for every Δ, and for every $B \in F^*$ we have:

$$P(B|A \cap PR(A, \Delta)) = P(B|A)$$

where $P(\ |\ )$ is the conditional probability. It means that B and $pr(A, \Delta)$ are independent with respect to $P(\ |A)$. A sufficient condition for this is that the field generated by F and by all the events $pr(A, \Delta)$, $A \in F$, and the field F^* are independent fields with respect to P.

Also Shafer's system of belief functions, (1976), can be viewed naturally from the point of higher order probabilities; the equality $Bel(A) = \alpha$ is translated into the event $pr(A, [\alpha, 1])$. However, in order to treat Dempster's combination rule, which constitutes the heart of this system, we shall have to use HOPs with many "incomparable" stages (experts). The additional notions and the precise independence assumptions which underlie Dempster's rule still need to be worked out.

A related line of further research is the study of HOPs in the context of the logic of knowledge – an area which has focused recently a great amount of activity; see Halpern and Moses (1985) and Halpern (1986) for an overview of this domain. The logic of HOPs, as outlined in the last section, is a natural probabilistic generalization of a logic of knowledge. This is apparent if, as Vaughan Pratt has suggested, we employ a notation of monadic operants: Rewrite '$pr_t(A, \Delta)$' as '$K_t^{\Delta}A$'. Our previous Alice and Bob example is now formalized as:

$$K_{\mathrm{Bob}}{}^{[x,x]}\, X \subset K_{\mathrm{Alice}}{}^{[0.7,1]}\, K_{\mathrm{Bob}}{}^{[x-0.05,\, x+0.05]}\, X$$

The usual knowledge operant K_t is the special case of K_t^{Δ}, where $\Delta = [1, 1]$. A study is called for concerning the notions, equalities and theorems which generalize from the modal-epistemic case to the probabilistic one.

Higher order probabilities have been considered by De Finetti, but rejected by him owing to his extreme subjectivist views. Savage considered the possibility but did not take it up, fearing that the higher order probabilities will reflect back on the ground level, leading to inconsistencies. Instances of higher order probabilities figure in works of Good (1965) and Jaynes (1958). More recent works are by Miller (1966), Gardenförs (1975) (for qualitative probabilities), Skyrms (1980a, b, 1985) – who did much to clarify matters, Goldstein (1983, 1985) and van Fraassen (1984). Also works by Lewis (1980) and van Fraassen (1980) presuppose a dualist framework based on the notion of objective chances. Such frameworks are special cases of HOPs, hence their relevance to the present work. Indeed, some versions of the principles to be considered here have been stated by Lewis, in a different conceptual setting. Other contributions to higher order probabilities have come from game theory and econometrics. Among the relevant works are those of Auman (1976), Armbuster and Böge (1979), Böge and Eisele (1979), and Mertens and Zamir (1985). Like the general subject of probability, our subject occupies a crossroads of several disciplines with diverse interests and different emphasizes. When I obtained the results of this work and wrote the first version of the paper I was unaware of much of this literature. The bibliography which has now been extended is still not intended as a complete list.

Due to limitations of space and deadline I have not entered into most of the proofs. I have preferred to give a summary overview, to be elaborated later in other papers. Some of the material has been abridged; I have included some illustrative examples of simple HOPs, but not the more interesting ones of general HOPs which arise naturally in distributed systems.

SIMPLE HOPS

Definition and Basic Properties

As in Kolmogoroff's framework (1933) we interpret propositions as subsets of some universal set, say W, and we refer to them as *events*.

We can regard W as the set of all possible worlds. Thus we have: X = set of all worlds in which X is true. Disjunctions conjunctions and negations become, respectively, unions, intersections and complements.

The possible world terminology which we shall use in this work does not mark any particular philosophical view. It is chosen for its convenience and visual suggestiveness. You can regard possible worlds merely as a formal tool; you can also, if you like, endow them with your favorite metaphysics.

Terminology: A *Boolean Algebra*, of subsets of W, is a class of subsets of W closed under the finitary Boolean operations. A *field*, known also as a σ-algebra, is a Boolean algebra closed under countable unions. The field (Boolean algebra) *generated by* a class S of sets is the smallest field (Boolean algebra) which contains S as a subclass. A field is *countably generated* if it has a countable set of generators. The probabilities are assumed here to be countably additive. The reader can easily see what the weaker, finitely additive version would be. Some of the results hold also for this version.

A HOP is a 4-tuple (W, F, P, pr), where F is a field of subsets of W, to be called *events*, P is a probability over F and pr is a mapping associating with every $A \in F$ and every real closed interval Δ an event $pr(A, \Delta)$.

$$pr\colon F \times \textit{set of closed intervals} \rightarrow F$$

As explained in the introduction, $pr(A, \Delta)$ is the event that the expert probability of A lies in Δ. P is the agent's subjective probability.

We put: $\mathbf{0}$ = empty subset of W = False, $\mathbf{1} = W$ = True, $\emptyset$ = empty interval (considered here as a closed interval). "Probability" is used occasionally both for the agent's and for the expert's probability; the intended reading is indicated by the context.

The following axioms are postulated for a HOP:

(I) $pr(A, [0, 1]) = pr(W, [1, 1]) = \mathbf{1}$ (The events that the probability of A lies in [0, 1] and that the probability of W is 1 are True.)

(II) $pr[A, \emptyset] = \mathbf{0}$ (The event that A's probability lies in $\emptyset$ is $\mathbf{0}$, i.e., False.)

(III) *If* $\Delta_1 \cup \Delta_2$ *is an interval then:*

$$pr(A, \Delta_1 \cup \Delta_2) = pr(A, \Delta_1) \cup pr(A, \Delta_2)$$

(I.e., A's probability lies in the interval $\Delta_1 \cup \Delta_2$ iff it lies either in Δ_1 or in Δ_2)
In the following two axioms "n" is a running index ranging over $\{1, 2, \ldots\}$.

(IV) $$\cap_n pr(A, \Delta_n) = pr(A, \cap_n \Delta_n)$$

(I.e., A's probability lies in every Δ_n iff it lies in their intersection).

(V) *If, for all* $n \neq m$, $A_n \cap A_m = \emptyset$, *then:*

$$\cap_n pr(A, [\alpha_n, \beta_n]) \subset pr(\cup_n \alpha_n, [\Sigma_n \alpha_n, \Sigma_n \beta_n])$$

(I.e., for pairwise disjoint A_n's, if A_n's probability lies in $[\alpha_n, \beta_n]$, $n = 1, 2, \ldots$, then the probability of $\cup_n A_n$ lies in $[\Sigma_n \alpha_n, \Sigma_n \beta_n]$. In the finitely additive version n ranges only over finitely many integers.)

Note that axioms (I)–(V) involve only W, F and pr. The crucial axiom which connects pr with P will be stated later.

THEOREM 1. *For every HOP* $H = (W, F, P, pr)$ *there is a mapping* p *which associates with every* $x \in W$ *a probability* p_x *over* F, *such that:*

(1) $$pr(A, \Delta) = \{x : p_x(A) \in \Delta\}$$

The mapping p *is uniquely determined by* (1) *and can be defined by:*

(2) $$p_x(A) = \inf\{\alpha : x \in pr(A, [0, \alpha])\}$$

as well as by:

(2′) $$p_x(A) = \sup\{\alpha : x \in pr(A, [\alpha, 1])\}.$$

Vice versa, if, for every $x \in W$, p_x *is a probability over* F *and* $\{x : p_x(A) \in \Delta\}$ *is in* F *for all* $A \in F$ *and all real closed* Δ, *then, if we use* (1) *as a definition of* pr, *Axioms* (I)–(V) *are satisfied.*

We call p the *kernel* of the HOP.

The proof of Theorem 1 is nothing more than a straightforward derivation of all the required details from the axioms, using (2) as the definition of p_x. (The *vice versa* part is even more immediate than the first part.)

We can give p_x an obvious meaning: It is the probability function of the world-x-expert.[2]

We can now extend pr and define $pr(A, \Xi)$ for arbitrary subsets Ξ of reals:

$$pr(A, \Xi) =_{\text{df}} \{x : p_x(A) \in \Xi\}.$$

If Ξ is a Borel set then $pr(A, \Xi)$ is in F. For $\alpha \in [0, 1]$, put: $pr(A, \alpha) =_{\text{df}} pr(A, \{\alpha\}) = \{x : p_x(A) = \alpha\}$. Evidently, for each fixed A, the set of all $pr(A, \Xi)$ is a field whose atoms are the $pr(A, \alpha)s$.

Define the HOP to be *discrete* if P is a discrete probability over F, i.e., some countable subset of W gets probability 1. In this case we can assume without loss of generality that W itself is countable and $\{x\} \in F$ for all $x \in W$. Define the HOP to be *pr-discrete* if, for each $A \in F$, P is discrete over the field of events of the form $pr(A, \Xi)$. This means that, for each A, there are countably many reals, α_n, $n = 1, 2, \ldots$, such that $\Sigma_n P(pr(A, \alpha_n)) = 1$. Evidently, if the HOP is discrete then it is *pr*-discrete; but not *vice versa*. If we restrict ourselves to *pr*-discrete HOPs we need not consider $pr(A, \Delta)$ in general, but only the special cases $pr(A, \alpha)$, where $\alpha \in [0, 1]$. Our five axioms are then equivalent to the following three:

$$\cup_\alpha pr(A, \alpha) = pr(W, 1) = \mathbf{1}$$

where α ranges over $[0, 1]$.

For $\alpha \neq \beta$,

$$pr(A, \alpha) \cap pr(A, \beta) = \mathbf{0}.$$

For pairwise disjoint $A_n s$,

$$\cap_n pr(A_n, \alpha_n) \subset pr(\cup_n A_n, \Sigma_n \alpha_n).$$

But the restriction to *pr*-discrete HOPs is quite severe. We *do* want to consider assignments which give some $pr(A, \Delta)$ a non-zero value, while assigning each $pr(A, \alpha)$, $\alpha \in \Delta$, the value 0. In order to do so we have to use $pr(A, \Delta)$, rather than $pr(A, \alpha)$ as a basic notion.

Notation: For $x, y \in W$, $A \in F$, put: $p(x, A) =_{\text{df}} p_x(A)$; assuming $\{y\} \in F$, put: $p(x, y) =_{\text{df}} p(x, \{y\})$ and $P(y) =_{\text{df}} P(\{y\})$.

If the HOP is discrete, then P is obviously determined by the values $P(x)$, $x \in W$. Thus, ordering W, we can represent P as a *probability vector* (a countable vector of non-negative entries which sum up to 1). Similarly the kernel p becomes a *probability matrix* (a countable square matrix in which every row is a probability vector).

Mathematically, any simple HOP amounts to a Markov process with initial probability P and transition probabilities $p(x, \ )$, $x \in W$. But the interpretation is altogether different and it motivates particular questions. The connection between P and the kernel p is determined by our sixth axiom.

Axiom VI

Let $P(A \mid B)$ be *the conditional probability of* A, *given* B. It is defined in the case that $P(B) \neq 0$ as $P(A \cap B)/P(B)$.

(VI) *If C is a finite intersection of events of the form $pr(B_i, \Delta_i)$ and if $P(pr(A, \Delta) \cap C) \neq 0$, then:*

$$P(A \mid pr(A, \Delta) \cap C) \in \Delta$$

The meaning of this axiom becomes clearer if we consider its particular case where C is omitted (i.e., C = empty intersection = W). This weaker form is:

($\mathrm{VI_w}$) *If $P(pr(A, \Delta)) \neq 0$, then:*

$$P(A \mid pr(A, \Delta)) \in \Delta$$

($\mathrm{VI_w}$) is the interval generalization of a principle suggested by Miller[3] (1966). Miller's Principle is obtained if Δ is a single point, it is in our notation: $P(A \mid pr(A, \alpha)) = \alpha$. Axiom ($\mathrm{VI_w}$) asserts the following: Given that in a more knowledgeable state my probability for A is in Δ, my conditional probability for A is in Δ. The plausibility of this rule derives from one's preference for probabilities of the more knowledgeable states. This intuition is borne out by proofs which derive Miller's principle and its generalizations from coherence: When probabilities are identified with certain parameters in gambling situations, it is shown that if the agent's probability function violates the principle then his chosen strategy will be incoherent; that is to say, another available strategy will do strictly better *under all possible outcomes*. Constructing a Dutch book, i.e., a system of bets in which the agent must lose irrespective of outcome, is such a coherence proof. Goldstein (1983) gave a coherence argument for ($\mathrm{VI_w}$). Actually, his stated principle is not ($\mathrm{VI_w}$) but something weaker (a condition equivalent to what we later designate by (3)), but in the course of the proof he derives ($\mathrm{VI_w}$). Independently, van Fraassen (1984) gave a simple Dutch book argument for Miller's principle. The underlying idea of the proofs can be used to derive the full Axiom (VI). Both proofs require that the expert be identified with the agent at some future time. We shall present in the coming subsection a more developed version in which this requirement is not necessary. Roughly speaking, what is needed is that *in principle, the expert probabilities be knowable to the agent in circumstances in which, knowing them, he will adopt them.* The precise meaning of this is given in the next subsection.

Axiom (VI) is the generalized interval form of a principle proposed by Skyrms (1980a) under the name "Generalized Miller". Skyrms' principle is the special case obtained when Δ and all $\Delta_i s$ are single points. A related principle has been proposed by Lewis (1980), under the name "Principal Principle".[4]

The restriction of C in Axiom (VI) to intersections of $pr(B, \Delta)$s is essential. For these are the events that describe the expert probability. The removal of the restriction on C will trivialize the HOP; for if C can be any event then one can easily show, by choosing C to be A, that (for almost all x) p_x takes only 0 and 1 as values, meaning that all facts are known to the expert.

As we shall see ($\mathrm{VI_w}$) is strictly weaker than (VI). One can however prove (using the results in the sequel) that we can restrict C to be a *single* event of the form $pr(B, \Delta)$; this apparently weaker form implies the full axiom.

It is not difficult to derive from each of the axioms (VI), ($\mathrm{VI_w}$) its more general version in which Δ can also be an open or half open interval. (Here the countable additivity of P is needed.) But Δ must be a convex set, otherwise the rule is not true in general. We shall see that (VI) implies a generalization to n-dimensional spaces, where Δ is replaced by some arbitrary n-dimensional convex set.

The Coherence Argument for Axiom (VI)

The coherence argument to be given here is a development of Goldstein's idea (1984). Van Fraassen's idea can be similarly used. We embody the argument in an example from which the general abstract form is easily seen.

Let A = the event that Silver Blaze will win next week's race. You, the agent, regard Mr. Holmes as an expert more knowledgeable than yourself and you are ready to replace your current probability, represented by the function $P(\ )$, by his. Holmes' probability is $P_H(\ )$, and $pr(B, \Delta)$ is the event that $P_H(B) \in \Delta$. Let

$$D = pr(A, \Delta) \cap C$$

where $C = pr(B_1, \Delta_1) \cap \ldots \cap pr(B_n, \Delta_n)$. Say $\Delta = [\alpha, \beta]$. Assuming that $P(D) > 0$, we have to show that

$$\alpha \leqslant P(A \mid D) \leqslant \beta$$

We shall derive $P(A \mid D) \leqslant \beta$. The other inequality is similarly proved. Assume $\beta < 1$, otherwise the claim is trivial. Let $\epsilon > 0$. Let *l*1 be a lottery ticket representing a bet on A with odds $\beta + \epsilon : 1$ and total payoff 1; i.e., *l*1 costs $\beta + \epsilon$ and pays 1 if Silver Blaze wins, 0 – otherwise. Let *l*2 be a ticket which costs $1 - \beta$ and which pays 1 if Silver Blaze does not win, 0 – otherwise. Your utility function for buying a ticket is a random variable whose value for each $w \in W$ is your utility which results from buying the ticket. Let X and Y be, respectively, the utility functions for *l*1 and *l*2. Identifying, as is customary, an event with its indicator function – the function which is 1 on the event and 0 on its complement – we get:

$$X = (1 - \beta - \epsilon) \cdot A - (\beta + \epsilon) \cdot (1 - A)$$
$$Y = \beta \cdot (1 - A) - (1 - \beta) \cdot A$$

Now, Holmes has locked in a drawer a note containing his probabilities for A and for $B_1, \ldots, B_n$, which drawer you are now attempting to open. Let *l*3 be a lottery ticket which costs and pays like *l*1, but which also carries with it the following option: If you succeed to unlock the drawer within the next half hour and if D is the case then you may at that moment exchange the bet represented by *l*1 for that represented by *l*2. Let Z be the utility function for *l*3.

Let $E(\ )$ be the expectation function for the probability $P(\ )$. Given a choice between two actions, you choose that of higher expected utility.[5] Evidently, *l*3 is at least as good as *l*1, because you can always choose to ignore the additional option. Coherence therefore implies:

$$E(X) \leqslant E(Z)$$

Let K be the event of your succeeding to unlock the drawer within the next half hour. The crucial point is that if K takes place you will adopt Holmes' probability for A and your expected values for X and for Y will at that moment be:

$$(1 - \beta - \epsilon) \cdot P_H(A) - (\beta + \epsilon) \cdot (1 - P_H(A))$$
$$\beta \cdot (1 - P_H(A)) - (1 - \beta) \cdot P_H(A)$$

If D is the case then $P_H(A) \leqslant \beta$ and this implies that the first value is negative and the second is not. Hence, in that situation you will prefer $l2$ to $l1$. Thus, if you buy $l3$ and if $K \cdot D$ is the case, you will realize the option. Since this option is available exactly when $K \cdot D$ is the case we get:

$$Z = (1 - K \cdot D) \cdot X + K \cdot D \cdot Y$$

This, together with $E(X) \leqslant E(Z)$, implies:

$$E(K \cdot D \cdot X) \leqslant E(K \cdot D \cdot Y)$$

Since $E(K \cdot D \cdot X) = E(X | K \cdot D) \cdot P(K \cdot D)$ and a similar equality holds for $E(K \cdot D \cdot Y)$ we get:

$$P(K \cdot D) \cdot \{E(X | K \cdot D) - E(Y | K \cdot D)\} \leqslant 0$$

Putting

$$E(X | K \cdot D) = (1 - \beta - \epsilon) \cdot P(A | K \cdot D) - (\beta + \epsilon)(1 - P(A | K \cdot D))$$

$$E(Y | K \cdot D) = \beta(1 - P(A | K \cdot D)) - (1 - \beta)P(A | K \cdot D))$$

we get:

$$P(K \cdot D) \cdot \{2P(A | K \cdot D) - 2\beta - \epsilon\} \leqslant 0$$

We assume for our argument that $P(K) > 0$. We also assume that K and A are probabilistically independent and so are K and $A \cdot D$, all with respect to your probability $P(\ )$; i.e., your success in unlocking the drawer has nothing to do with the race's outcome or with the contents of Holmes' note. These assumptions imply that $P(K \cdot D) = P(K) \cdot P(D) > 0$ and that $P(A | K \cdot D) = P(A | D)$. Consequently we get: $P(A | D) \leqslant \beta + \epsilon/2$. Since ϵ is arbitrary we get the desired inequality.

Note that our argument requires that your unlocking the drawer within the next half hour will not destroy the agent–expert relation that you bear to Holmes; whatever you have learned by that time, you will still adopt Holmes' probability for A the moment you see it. If we have some doubts about this we can replace "half hour" by "five minutes"; as long as $P(K) > 0$ the argument will work. Furthermore, it will work if your adoption of Holmes' probability can be guaranteed with sufficiently high probability. All in all, we need the following assumptions in the general setting:

For any events A, $B_1, \ldots, B_n$ and any real closed intervals Δ, Δ_1, $\ldots$, Δ_n, there exists an event K of non-zero probability such that:

(i) If K occurs then, at that moment, the agent will know and adopt the expert probability for A.

(ii) K is independent of A as well as of $A \cap D$, where D is defined as above.

Actually a weaker condition will do: For every $\delta > 0$, there exists a K of non-zero probability satisfying (ii) such that, given K, the conditional probability of the agent adopting the expert probability for A is $>1 - \delta$. If our field of events does not contain such a K, we require that it be possible to enlarge it by adding such an event. The operational meaning of expert probability is given through this requirement. In the example K appeared as "the unlocking of the drawer". In the abstract setting K constitutes the possibility of knowing the expert probabilities for the given events in an independent way which does not destroy the agent–expert relation.

Our argument does not require that the expert be identified with the future time agent. But we still need to associate with certain events a temporal parameter – the time of their occurrence. For a complete formalization of the argument we have to introduce time explicitly into the framework. Events are to be represented as pairs (B, τ), where $B \subset W$ and τ is the time associated with the event's occurrence; the probability of the event is our old $P(B)$, but events with different τ's are different. If the expert is identified with the agent at some future time, say tomorrow, then our K is simply the event of "its being tomorrow"; it is represented as (W, *tomorrow*). In that case $P(K) = 1$ and all the needed assumptions are trivially satisfied.

The Main Consequences of Axiom (VI)

In what follows integrating a function $f(t)$ with respect to a probability m is written as $\int f(t) \cdot m(dt)$.

LEMMA 1. *Axiom* ($\mathrm{VI_w}$) *implies that the following holds for all* $A \in F$:

$$(3) \qquad P(A) = \int p(x, A) \cdot P(dx)$$

The proof consists in applying the formula $P(A) = \Sigma_i P(A \mid B_i) \cdot P(B_i)$, where the B_i's form a partition, passing to the limit and using the definition of an integral.

The implication (3) $\Rightarrow$ (VI_w) is not true in general. Note that in the discrete case (3) becomes:

$$(3_d) \qquad P(x) = \Sigma_y p(x, y) \cdot P(y)$$

which means that the probability vector is an eigenvector of the kernel.

Terminology: A *0-set* is a set of probability 0. Something is said to hold for *almost all* x if it holds for all x except for a 0-set. The probability in question is P, unless specified otherwise.

Definition (i) If S is a class of events then $pr[S] =_{df}$ the field generated by all events of the form $pr(A, \Delta)$, where A ranges over S and Δ over the real closed intervals.

(ii) Two worlds $x, y \in W$ are *epistemically equivalent over* S, (or, for short, *equivalent over* S) if $p_x(A) = p_y(A)$ for all $A \in S$. If $S = F$ (the entire field), we say that x and y are *epistemically equivalent* (or, for short, *equivalent*); this simply means that $p_x = p_y$.

(iii) $[x]_S$ is the equivalence-over-S class containing the world x. If $S = F$ we omit the subscript and write simply $[x]$.

It is easily seen that x and y are equivalent over S iff, for all $A \in pr[S]$, $x \in A \Leftrightarrow y \in A$. If $pr[S]$ is generated by countably many generators, say X_n, $n = 1, 2, \ldots$, then it is easy to show that the equivalence classes over S are exactly all non-empty intersections $\cap X'_n$ where each X'_n is either X_n or its complement. Hence the equivalence classes are themselves in $pr[S]$, they are exactly the *atoms* of this field. Hence, if $pr[S]$ is countably generated, then every $[x]_S$ is in it and, afortiori, is in F. The following lemma shows that the condition that $pr[S]$ be countably generated is rather mild.

LEMMA 2. *If S is either countable or a countably generated field then $pr[S]$ is countably generated.*

The first case follows from the observation that, in the definition of $pr[S]$, we can let Δ range only over rational intervals, i.e., intervals with rational end points. For the second case, let S be a field generated by a countable set of events and let S' be the Boolean algebra generated by the same set. Then S' is countable and the claim follows by showing that $pr[S] = pr[S']$.

THEOREM 2. *Axiom* (VI) *is equivalent to the conjunction of* (3) *with any of the following conditions:*

(A) *For each* $A \in F$ *and each* Δ, *for almost all* x, $x \in pr(A, \Delta) \Rightarrow p_x(pr(A, \Delta)) = 1$. (*Or, equivalently,* $pr(A, \Delta) - pr(pr(A, \Delta), 1)$ *is a* 0-*set.*)

(B) *For each countably generated field* $S \subset F$, *for almost all* x, $p_x([x]_S) = 1$.

(C) *For each* $A \in F$, *for almost all* x, $p_x(A) = \int p_y(A) \cdot p_x(dy)$.

Here I shall only show how to get (A) from Axiom (VI). Let $0 \leqslant \beta \leqslant 1$. Let

$$D = pr(pr(A, \Delta), [0, \beta]) \cap pr(A, \Delta)$$

From Axiom (VI) (with $pr(A, \Delta)$ in the role of A, $[0, \beta]$ in the role of Δ and the "second" $pr(A, \Delta)$ in the role of C) we get: If $P(D) > 0$ then $P(pr(A, \Delta) | D) \leqslant \beta$. But $D \subset pr(A, \Delta)$, hence $P(D) > 0$ implies $P(pr(A, \Delta) | D) = 1$. Therefore, whenever $\beta < 1$ we must have $P(D) = 0$. This implies that almost all of $pr(A, \Delta)$ is included in $pr(pr(A, \Delta), 1)$.

The implication (A) $\Rightarrow$ (B) follows, via standard measure theory, from the basic properties of the equivalence classes $[x]_S$. Also the implications (B) $\Rightarrow$ (C) and (3) + (B) $\Rightarrow$ (VI) do not require more than standard measure theory. The present derivation of (B) from (3) + (C) uses advanced ergodic theory[6] and I do not know if this can be avoided. Fortunately the rest of this paper does not rely on this implication (except the corresponding implication for general HOPs). Note that in the discrete case (C) is equivalent to:

(C_d) $\quad p(x, z) = \Sigma_y p(x, y) \cdot p(y, z)$

It means that the kernel-matrix is equal to its square.

(A) says the following: If the expert probability assigns an event A a value in Δ, then it assigns the value 1 to the event that the expert probability of A is in Δ. If we picture our expert as a conscious being, then (A) can be regarded as an introspection axiom: The expert knows for sure the values he assigns. This would make (A) very natural, an almost "self-evident" requirement.[7] But this intuition rests on our interpreting events of the form $pr(A, \Delta)$ in terms of some conscious being. What if the expert probability is given through some preprogrammed computer? It would be up to the programmer to decide whether the program should satisfy (A). Or what if the expert probability is interpreted as objective chance?[8] The derivation of (A) from (VI), hence – from coherence – obviates speculations about the expert. The precise meaning of "expert probability" consists in its operational implications for the agent, as described in the preceding section and (A) must be satisfied if we want our *agent* to be coherent.

If the field F of the HOP is countably generated (a very mild assumption) then (B) is equivalent to:

(B′) *For almost all x* $p_x([x]) = 1$
(Because, for all $S \subset F$, $[x] \subset [x]_S$)

Now (3) + (B′) means that we get a *disintegration* of the probability P, in which the epistemic equivalence classes are the fibers. The picture is as follows:

Let $\{E_u : u \in U\}$ be the family of epistemic equivalence classes, with different indices marking different classes. Let P_u be the common p_x for $x \in E_u$. The class of all $V \subset U$ such that $\cup_{u \in U} E_u \in F$, forms a field. Define over it a probability m by:

$$m(V) =_{\text{df}} P(\cup_{u \in V} E_u)$$

Then, (3) + (B′) is equivalent the following condition.

(D) *For all A in F*

$$P(A) = \int_U P_u(A) \cdot m(du)$$

and for almost all (with respect to m) u

$$P_u(E_u) = 1$$

If W is countable the disintegration means the following: After deleting from the kernel-matrix rows and columns which correspond to some set of probability 0, the rest decomposes into submatrices around the main diagonal in each of which all rows are equal; P itself is a mixture of these rows. Such HOPs are exactly those that can be constructed as follows:

- *Choose a partition* $\{E_u : u \in U\}$ *of W into non-empty disjoint sets, with different indices marking different sets.*
- *Choose for each u in U a probability*, P_u, *on W such that the set U′ consisting of all u such that* $P_u(E_u) = 1$ *is not empty.*

- *Choose a probability, m, on U such that $m(U') = 1$, and let P be the mixture of the P_u's with weight function m (i.e., P is defined by the first equality in* (D)).
- *For each $u \in U$ and each $x \in E_u$ put $p_x = P_u$ and define $pr(A, \Delta)$ to be $\{x : p_x(A) \in \Delta\}$.*

The construction is essentially the same for a general countably generated F; some additional stipulations of measurability should be included in order to make possible the formation of the mixture and to ensure that the $pr(A, \Delta)$'s are in F.

Call P_u and its corresponding equivalence class, E_u, *self-supporting* if $P_u(E_u) = 1$; call it and its corresponding class *coherent* if P_u is a mixture of self-supporting P_u's. Call a world self-supporting (coherent) if its equivalence class is. A self-supporting class is of course coherent. A coherent class which is not self-supporting must get the value 0 under its own P_u. The union of all self-supporting classes is a set of probability 1. As far as probabilities are concerned we can restrict ourselves to this part of the HOP. But it is sometimes convenient to use coherent non self-supporting worlds as representatives of states of knowledge in which the agent knows for sure that his current probability function is not the expert probability. Non self-supporting worlds may also be significant in modal interpretations of HOPs which distinguish between the empty set and 0-sets.

Examples

Example 1: $W = \{w1, w2, w3\}$ $P = \{1/3, 1/3, 1/3)$ and the kernel matrix is:

0.5	0.5	0.0
0.0	0.5	0.5
0.5	0.0	0.5

The agent's current probability assigns each world the value 1/3. In world $w1$ the expert knows that he is not in $w3$ and he assigns each of $w1$, $w2$ the value 0.5. Similarly, in $w2$ the expert assigns 0.5 to each of $w2$, $w3$, etc. By direct checking one can verify that ($\mathrm{VI_w}$) is satisfied. The checking here is easy because $pr(A, \alpha) \neq \emptyset$ only for $\alpha = 0, 0.5, 1$. Since, for $i \neq j$, $p_{wi} \neq p_{wj}$, each $\{wi\}$ constitutes an epistemic equivalence class; and since $p_{wi}(\{wi\}) = 0.5$ none of the classes is self-supporting. Hence Axiom (VI) is violated. Indeed, writing "wi" for "$\{wi\}$", we have: $P(w1 \mid pr(w1, 0.5) \cap pr(w2, 0.5)) = P(w1 \mid w1) = 1 \neq 0.5$. The anomaly in this case is expressible as follows: The probabilities of the $w1$-expert for $w1$ and for $w2$ imply that our world is $w1$, which is more than the expert himself knows. ($\mathrm{VI_w}$) is therefore strictly weaker than (VI). The following 2-world example satisfies (3) but violates ($\mathrm{VI_w}$). The matrix is:

$$\begin{array}{ll} 0.8 & 0.2 \\ 0.2 & 0.8 \end{array}$$

and the vector is (0.5, 0.5).

Example 2 $W = \{w1, w2, \ldots, w8\}$, P is:
(0.1, 0.2, 0.2, 0.1, 0.4, 0, 0, 0) and the kernel matrix is:

$$\begin{array}{llllllll} 0.2 & 0.4 & 0.4 & & & & & \\ 0.2 & 0.4 & 0.4 & & & & & \\ 0.2 & 0.4 & 0.4 & & & & & \\ & & & 0.2 & 0.8 & & & \\ & & & 0.2 & 0.8 & & & \\ & & & & & 1 & & \\ 0.05 & 0.2 & 0.2 & 0.05 & 0.2 & 0.5 & 0.0 & 0.0 \\ 0.2 & 0.1 & 0.1 & 0.2 & 0.1 & 0.1 & 0.1 & 0.1 \end{array}$$

where all undisplayed entries are 0. The sets $\{w1, w2, w3\}$, $\{w4, w5\}$ and $\{w6\}$ are equivalence classes which are self-supporting. P is a mixture of these 3 types of rows, with weights 0.5, 0.5, 0, respectively. Hence (D) is satisfied, therefore also Axiom (VI). $w7$ is a coherent non-self-supporting world, because the 7th row is a mixture of the first three types (with weights 0.25, 0.25, 0.5). $w8$ is not coherent.

The example can be provided with the following scenario. A number is to be chosen from $\{1, 2, 3\}$. For $i = 1, 2, 3$, the number chosen in wi is i, and in all of these 3 worlds the expert assigns the probabilities, 0.2, 0.4, 0.4, to the outcomes 1, 2, 3, respectively. In $w4$ the number chosen is 1 and in $w5$ it is 2; and the expert probabilities in both are 0.2, 0.8. In $w6$ the number is 2 and it is also assigned probability 1. In the agent's current state he assigns probability 0 to the third state of expert knowledge, and equal probabilities to the first and second states. World $w7$ represent a similar situation but with different weights. We can imagine 3 lotteries for choosing the number; in each equivalence class the expert knowledge is knowledge of the chosen lottery but not of the actual outcome (except for the third lottery in which the outcome is fixed).

Example 4. Let H be the probability of "heads" of some given coin of unknown bias. Treat H as a random variable. The agent's knowledge is represented by a probability distribution for H. Say it is the uniform distribution over [0, 1]. The expert does not know the value of H but he has some additional information. Say his additional information is the value of N – the number of "heads" in 50 independent tosses. Then our worlds can be regarded as pairs (h, n), where h is a real number in [0, 1] and n is an integer between 0 and 50; (h, n) is the world in which the two equalities $H = h$ and $N = n$ are true. The field F is generated by the sets of the form $[\alpha, \beta] \times \{n\}$, where $0 \leqslant \alpha \leqslant \beta \leqslant 1$, and $n = 0, \ldots, 50$.

Given $H = h$, we get the binomial distribution $b_{h,50}$ for N. This fact, together with the agent's uniform distribution for H, determines his probability P:

$$P(A) = \int_0^1 b_{h,50}(A) \cdot \mathrm{d}h$$

The expert's probability in world (h, n) is obtained by conditioning on his information, it is $P(\;|N = n)$. There are 51 equivalence classes which correspond to the 51 possible values of N and all worlds are self-supporting.

Let $\alpha(n) = P(H \leqslant 0.5 | N = n)$. Then it can be shown that $\alpha(n)$ is strictly decreasing as n increases. Hence $\alpha(n)$ determines uniquely n; it is easily seen that this means that the event $N = n$ is equal to $pr(H \leqslant 0.5, \alpha(n))$. The whole field F is therefore generated by events which are either of the form $H \in [\alpha, \beta]$ or of the form $pr(H \in [\alpha, \beta], \Delta)$. Consequently we can give an abstract description of this HOP which does not mention the fifty tosses. The only function of the tosses is to affect the distribution of H; in our framework such changes in distribution constitute themselves events which can be treated directly, without having to bring in their causes.

Sub-HOPs and Iterated pr

Definition. Given a simple HOP (W, F, P, pr) and given $S \subset F$, define $H[S]$ as the smallest field containing S and closed under *pr* (i.e., $A \in H[S] \Rightarrow pr(A, \Delta) \in H[S]$ for every real closed interval Δ).

The structure consisting of W, $H[S]$ and the restrictions of P and *pr* to $H[S]$ is *the sub-HOP generated by S*

LEMMA 3. *If S is a Boolean algebra and, for every A in S and every rational closed interval* Δ, $pr(A, \Delta)$ *is in S, then* $H[S]$ *is the field generated by S (i.e., generated using only the countable Boolean operations, not the operants* $pr(\ ,\Delta)$*).*

COROLLARY. *If S is either countable, or a countably generated field, then* $H[S]$ *is countably generated.*

This corollary shows that the assumption that the HOP has a countably generated field is rather mild. For, starting with any countably generated subfield of F, we can embed it in a sub-HOP with a countably generated field.

Say that A *is equal a.a. to* B if $A - B$ and $B - A$ are 0-sets. Say that two classes of sets are *equal modulo 0-sets* if every member of one is equal a.a. to some member of the other.

From Theorem 2 the following can be derived, assuming Axiom (VI):

THEOREM 3. *Let S be any subset of our field F. Then*

(i) *The fields* $pr[S]$, $pr[pr[S]]$ *and* $pr[H[S]]$ *are equal modulo 0-sets.*

(ii) *If S is a Boolean algebra then* $H[S]$ *is equal modulo 0-sets to the field generated by* $S \cup pr[S]$.

Note that, in general, for a subfield $G \subset F$, non of the fields G and $pr[G]$ includes the other. If G is closed under $pr(\ ,\ )$ then $pr[G] \subset G$; in general, this inclusion is proper.

Roughly speaking, (ii) means that if we disregard 0-sets we do not need to iterate *pr* in a nested way; for, modulo 0-sets, nested applications of *pr* reduce to non-nested applications. A related syntactical result is given in the last section.

Random Variables

Events can be regarded as a special kind of random variables (i.e., measurable real-valued functions); namely – those which take only the values 0 and 1. In generalizing to random variables the event's probability is replaced by the variable's expected value.

For a HOP (W, F, P, pr) and for a random variable Y, let $E(Y)$ and $e_x(Y)$ be the expected values of Y, for the probabilities P and p_x, respectively. Let $ex(Y, \Delta)$ be the event $\{x : e_x(Y) \in \Delta\}$. Then the generalization of $(\mathrm{VI_w})$ to random variables is:

$$(\mathrm{R_w}) \qquad E(Y \mid ex(Y, \Delta)) \in \Delta$$

This is the generalized interval form of a principle suggested by Skyrms (1985) in a discrete setting; he guessed that the principle is equivalent to his "Generalized Miller". Our general conjecture, of which we feel rather confident is:

$$\text{Axiom (VI)} \Leftrightarrow (\mathrm{R_w})$$

We have proved it for the discrete case and also the direction (VI) $\Rightarrow (\mathrm{R_w})$ in the general case.

The following can be considered as the many dimensional generalization of $(\mathrm{R_w})$:

(R) *For all n, if Y is a vector-valued random variable taking values in an n-dimensional Euclidean space and Λ is any-convex subset, then*

$$E(Y \mid ex(Y, \Lambda)) \in \Lambda$$

THEOREM 4. *Axiom* (VI) $\Leftrightarrow$ (R).

GENERAL HOPS

In general, a HOP is a structure of the form:

$$(W, F, P, T, pr)$$

where, as before, (W, F, P) is a probability space, $T = (T, \leqslant)$ is a partially ordered set and where

$$pr : F \times T \times \textit{set of closed intervals} \rightarrow F$$

$pr(A, t, \Delta)$ is the event that the probability of A at stage t lies in Δ. The partial ordering reflects a knowledge hierarchy: if $s \leqslant t$ then the knowledge at stage s is included in the knowledge at stage t. As before,

P is the current subjective probability; here "current" is earlier, i.e., less than or equally informative, than every stage in T. Put:

$$pr_t(A, \Delta) =_{\mathrm{df}} pr(A, t, \Delta)$$

For general HOPs the first five axioms (I*)–(V*) are the obvious generalizations of our previous axioms (I)–(V). Namely, we replace "pr" by "pr_t" and require that the condition hold for all $t \in T$.

Theorem 1 generalizes in the obvious way and we get, for each $t \in T$ and each $x \in W$, a probability $p_{t,x}$ which is related to pr_t by:

$$x \in pr_t(A, \Delta) \Leftrightarrow p_{t,x}(A) \in \Delta$$

$p_{t,x}$ represents *the stage-t-expert knowledge in the world x.*

The correct generalization of Axiom (VI) is not as obvious, but is not difficult to find:

(VI*) *For each $t \in T$, if C is a finite intersection of events of the form $pr_{s_i}(B_i, \Delta_i)$ where every s_i is $\leqslant t$, and if $P(pr_t(A, \Delta) \cap C)) \neq 0$, then:*

$$P(A \mid pr_t(A, \Delta) \cap C) \in \Delta$$

This axiom can be derived by the same kind of coherence argument which we used for Axiom (VI). The essential point is that, for given events, the agent can *in principle* know the expert probabilities of various stages, so that if one of the stages dominates the rest he will adopt the probability of that stage. Note that we make no stipulation if, for some s_i, $s_i \nleqslant t$. In particular, we leave open the question how one is to combine two probabilities for the same event supplied by two "incomparable" experts; I do not believe that a general rule can cover this case. But an analysis of various subcases is extremely important. (For example, when the two experts are in a certain sense "independent", then Dempster's rule can be used.)

As before, we consider fields generated by events of the form $pr_t(A, \Delta)$ and define epistemic equivalence. But now these concepts depend on the additional stage parameter which is displayed as an additional subscript. Thus $[x]_s$ is the epistemic equivalence class of world x at stage s, and, for a class S of events, $[x]_{s,S}$ the equivalence-over-S class at stage s; i.e., $y \in [x]_{s,S}$ iff $p_{s,y}(A) = p_{s,x}(A)$ for all $A \in S$. Each of the conditions (3), (A), (B) and (C) has a starred version which is the corresponding generalized condition for general HOPs. These general conditions are:

(3*) *If* $t \in T$ *and* $A \in F$, *then* $P(A) = \int p_{t,x}(A) \cdot P(\mathrm{d}x)$.
(A*) *If* $s \leqslant t$ *and* $A \in F$, *then, for almost all* $x \in pr_s(A, \Delta)$, $p_{t,x}(pr_s(A, \Delta)) = 1$. (*Or, equivalently,* $pr_s(A, \Delta) - pr_t(pr_s(A, \Delta), 1)$ *is a* 0-*set.*)
(B*) *If* $s \leqslant t$, S *is a countably generated subfield and* $A \in F$, *then* $p_{t,x}([x]_{s,S}) = 1$, *for almost all* x.
(C*) *If* $s \leqslant t$ *and* $A \in F$, *then, for almost all* x, $p_{s,x}(A) = \int p_{t,y}(A) \cdot p_{s,x}(\mathrm{d}y)$.

Theorem 2 for general HOPs is:

THEOREM (2*). *Each of the conditions* (3*) + (A*), (3*) + (B*) *and* (3*) + (C*) *is equivalent to Axiom* (VI*).

In the following corollary we assume for simplicity that the field of the HOP is countably generated.

COROLLARY. *The general HOP* (W, F, P, T, pr) *satisfies Axiom* (VI*) *iff the following two conditions hold:* (i) *For every* $t \in T$, (W, F, P, pr_t) *satisfies Axiom* (VI) *and* (ii) *For every* $s \leqslant t$, *for almost all* x, $(W, F, p_{s,x}, pr_t)$ *satisfies Axiom* (VI).

Further insight into the structure of general HOPs is given by the following theorem. Again we assume, for simplicity, that the field is countably generated. We use "$\simeq_t$" to denote epistemic equivalence at stage t:

$$x \simeq_t y \Leftrightarrow_{\mathrm{df}} p_{t,x} = p_{t,y}$$

THEOREM 5. *Axiom* (VI*) *holds iff for each* $t \in T$ *the simple HOP* (W, F, P, pr_t) *satisfies Axiom* (VI) *and the following holds:*

(E) *If* $s \leqslant t$ *then:* $x \simeq_t y \Rightarrow x \simeq_s y$ *for almost all* x, y (*i.e., for some* W' *of probability* 1 *this holds for all* $x, y \in W'$.)

(E) means that, as we pass to more progressive stages, almost everywhere epistemic equivalence is the same or becomes stronger; the partition into equivalence classes can change only by becoming more refined.

PROBABILITY LOGIC

We shall first consider the logic which corresponds to simple HOPs

Let Ξ be a set of reals such that $0, 1 \in \Xi$. Call an interval with end-points in Ξ a *Ξ-interval*.

Let PRL_Ξ be the calculus obtained by adjoining to the propositional calculus the following one-place sentential operants (or connectives): For each closed Ξ-interval, Δ, the one-place operant K^Δ. Given some class $\{X_i : i \in I\}$ of sentential variables, the class of all wffs (well formed formulas) of PRL_Ξ is the smallest class such that:

- Every sentential variable is a wff.
- If ϕ and ψ are wffs, so are $\neg\phi$ and $\phi * \psi$, where $*$ is any of the standard binary connectives.
- If ϕ is a wff and Δ is a closed Ξ-interval then $K^\Delta\,\phi$ is a wff.

We use "K" in order to accommodate our notation to the one used in logics of knowledge. We can read $K^\Delta\phi$ as: The expert knows that ϕ's probability lies in Δ.

Let $\mathbf{H} = (W, F, P, pr)$ be a simple HOP and let τ be an assignment of members of F to the sentential variables. Then the value, $|\phi|_{\mathbf{H},\tau}$, of the wff ϕ under τ is defined by:

- $|X|_{\mathbf{H},\tau} = \tau(X)$ for each sentential variable X.
- $|\neg\phi|_{\mathbf{H},\tau} = W - |\phi|_{\mathbf{H},\tau}$ $|\phi \vee \psi|_{\mathbf{H},\tau} = |\phi|_{\mathbf{H},\tau} \cup |\psi|_{\mathbf{H},\tau}$ and similarly for the other Boolean connectives.
- $|K^\Delta\phi|_{\mathbf{H},\tau} = pr(|\phi|_{\mathbf{H},\tau}, \Delta)$

Definition. A wff ϕ is *p-valid*, to be denoted: $\models_p \phi$, if, for every simple HOP $\mathbf{H}$ which satisfies Axiom (VI) and every τ, the probability of $|\phi|_{\mathbf{H},\tau}$ is 1. Two wffs ϕ, ψ are *p-equivalent* if $\phi \leftrightarrow \psi$ is *p*-valid.

Call ϕ *K-free* if it does not contain any K^Δ, i.e., it is a formula of the propositional calculus.

THEOREM 7. *Every wff of* PRL_Ξ *is p-equivalent to a Boolean combination of K-free formulas and formulas of the form* $K^\Delta\,\sigma$ *in which* σ *is K-free.*

This means that as far as probabilities are concerned (i.e., if we disregard 0-sets) we need not use nested K^Δs.

THEOREM 8. *Translate into* PRL_Ξ *the wffs of propositional modal logic with the knowledge (or necessity) operant K, by replacing each* $K\psi$ *by* $K^{[1,1]}\psi$. *Let* ϕ^* *be the translation of* ϕ. *Then:*

$$S5 \vdash \phi \quad \textit{iff} \quad \models_p \phi^*$$

Thus, the usual knowledge operant K can be construed as $K^{[1,1]}$ and $S5$ thereby becomes a fragment of PRL_Ξ.

It can be shown that, for Ξ = set of rational numbers, the set of p-valid wffs is recursive. Also PRL_Ξ can be provided with a natural set of formal axioms so that, with *modus ponens* as derivation rule, p-validity coincides with provability. We shall not dwell here on these developments; we hope to include them in one of the more elaborate papers on the subject.

Some Questions

Other validity notions can be considered (e.g., that $|\phi|_{\mathbf{H},\tau}$ always contains all coherent worlds in the HOP), as well as other interpretations of the necessity operant (e.g., as $\phi \wedge pr(\phi, [1, 1])$). What modal logics are thereby obtained?

The Logic of General HOPs

Fix a partially ordered set $T = (T, \leqslant)$. Let $\mathrm{PRL}_{\Xi,T}$ be the logic which corresponds to general HOPs with set of stages T. It is defined in the same way as PRL_Ξ, but the operants are K_t^Δ, where $t \in T$ and Δ is a closed Ξ-interval. The value $|\phi|_{\mathbf{H},\tau}$ of the wff ϕ in the general HOP $\mathbf{H}$, under the assignment τ, is defined in the obvious way by interpreting K_t^Δ as $pr_t(\ ,\Delta)$; p-validity and p-equivalence are defined accordingly.

Now consider a propositional modal language, $\mathbf{K}_T$, in which we have, instead of a single knowledge operant, an indexed family K_t, $t \in T$. $K_t\phi$ states that ϕ *is known at (or by) stage t*. The stages can be identified with experts, or – as is customary in the setting of logics of knowledge – with players, or processors; $s \leqslant t$ means that t knows at least as much as s.

As before we can translate from our modal language into $\mathrm{PRL}_{\Xi,T}$: For $\phi \in \mathbf{K}_T$, let ϕ^* be the wff obtained by replacing each $K_t\psi$ by $K_t^{[1,1]}\psi$.

THEOREM 9. *The set of all* ϕ *in* $\mathbf{K}_T$ *such that* ϕ^* *is p-valid is exactly the set of wffs derivable, by modus ponens and the rule: if* $\vdash \phi$ *then* $\vdash K_t\phi$, *from the following axioms:*

(i) *All tautologies.*
(ii) *For each* $t \in T$, *the axiom schemas of S5 – with K replaced by* K_t.
(iii) $K_s\phi \to K_t\phi$, *for each* $s \leqslant t$.

If the partial ordering $\leqslant$ is empty, then we are in the many-players situation without presuppositions about knowledge inclusions. In that case all the axioms reduce to (i) and (ii) and we get $S5_{(m)}$ – the many-player version of *S*5, cf. Halpern and Moses (1985). Thus we have:

COROLLARY $S5_{(m)} \vdash \phi$ *iff* $|=_p \phi^*$.

If Ξ = set of rationals and *T* with its partial ordering is given effectively, then the question of deciding the *p*-validity of wffs of $\mathrm{PRL}_{\Xi,T}$ arises. Since every wff involves only finitely many stages, the question is whether for finite *T*'s *p*-validity is decidable by a procedure which is uniform in *T*. I conjecture that the answer to this is positive. I also conjecture that there is a natural set of axioms, analogous to the axioms mentioned above for PRL_Ξ.

DIRECTIONS FOR FURTHER RESEARCH

Various notions arising in the framework of the logic of knowledge can be now carried over to the framework of HOPs. It is obvious, for example, what *common knowledge* will be in a HOP. To find out the probabilistic generalizations of these notions, their basic properties and the theorems which govern them is a natural line of research to follow.

Department of Mathematics
Hebrew University of Jerusalem, and
Stanford Research Institute

NOTES

[1] This extended preliminary report summarizes the current state of the work. Its first stage was described at a lecture given at the NSF Conference on The Foundations of Probability and Causality, UC Irvine July 1985. A later, more developed version was reported at the Conference on Theoretical Aspects of Reasoning about Knowledge.

(Sponsored by IBM. Am. Ass. of AI, Office of Naval Research and Ass. for Com. Mach.) Monterey, March 1986. Since then it has been further extended by various results. We preferred to give here an overview of the results leaving the details of the proofs and further discussions and analysis for future, more elaborate papers. We are indebted to Bill Harper, Richard Jeffrey and Brian Skyrms for useful comments and observations and to Isaac Levi and Amos Tversky for comments on an earlier draft.

[2] A similar function serves as a starting point in a different conceptual setting: Adopting a realistic metaphysics of possible worlds as well as of objective chances, Lewis (1980) correlates with every possible world the objective chance distribution in that world.

[3] Miller did not believe in the principle and used it in an unsuccessful *reductio ad absurdum* of the notion of higher order probabilities.

[4] In that principle C can be the whole world history up to the given time, conjuncted with the complete theory of objective chance which is valid in that world; thus it presupposes Lewis' metaphysics of possible worlds and objective chances.

[5] All that we need of the maximal utility principle are its following two implications: Let B be some event. (i) If you prefer (in the weak sense which includes being indifferent) $(1 - B)\cdot X + B\cdot Y$ to $(1 - B)\cdot X + B\cdot X$, then you should prefer $B\cdot Y$ to $B\cdot X$. (ii) If you prefer a gamble on *not-A* which costs $1 - \beta$ and pays 1 – all this conditonal on B, to a gamble on A which costs $\beta + \epsilon$ and pays 1 – also conditional on B, and if your probability of B is non-zero, then your conditional probability of A given B should be $\leqslant \beta + \epsilon/2$. It can be shown that if (ii) is violated then a Dutch book against you is obtainable, by replacing, in a system of bets which accord with your probabilities, one bet by another bet preferred to it.

[6] I am thankful to my colleagues at the Hebrew University H. Furstenberg, I. Katzenelson and B. Weiss for their help in this item. Needless to say that errors, if any, are my responsibility.

[7] So much so that Goldstein (1985) used implicitly an assumption equivalent to (A) in a crucial step in a proof (the passing from the second equality to the third, in the proof of his first theorem, p. 234), without remarking on its being an additional assumption.

[8] Indeed, a related thesis appears in Lewis (1980), where it is claimed that the numeric values of the chances are themselves no longer a matter of chance. The claim rests on his strong Principal Principle and involves his metaphysics of possible worlds and chances.

REFERENCES

Armbuster, W. and W. Böge (1979) 'Bayesian game theory', in *Game Theory and Related Topics*, O. Moeschlin and D. Pallaschke (eds.), North Holland, pp. 17–28.

Auman, R. J. (1976) 'Agreeing to disagree', *The Annals of Statistics* **4**(6), pp. 1236–1239.

Böge, W. and T. H. Eisele (1979) 'On solution of Bayesian games', *International Journal of Game Theory* **8**(4), pp. 193–215.

Dawid, A. P. (1982) 'The well-calibrated Bayesian', *Journal of the Am. Stat. Assoc.* **77**, pp. 605–613.

Domotor, Z. (1981) 'Higher order probabilities' (manuscript).

Gaifman, H. (1983) 'Towards a unified concept of probability', manuscript of invited lecture to the 1983 International Congress for Logic Philosophy and Methodology of

Science, Salzburg. Also in the proceedings: *Logic Methodology and Philosophy of Science VII*, R. Barcan Marcus G. J. W. Dorn and P. Winegartner (eds.) (Amsterdam: North-Holland 1986), pp. 319–350.

Gardenförs, P. (1975) 'Qualitative probability as intentional logic', *Journal of Philosophical Logic* **4**, pp. 171–185.

Good, I. J. (1965) *The Estimation of Probabilities* (Cambridge Massachusetts).

Goldstein, M. (1983) 'The prevision of a prevision', *Journal of the Am. Stat. Assoc.* **78**, pp. 817–819.

Goldstein, M. (1985) 'Temporal coherence', in *Bayesian Statistics 2*, J. M. Bernardo, M. H. DeGroot, D. V. Lindley and A. F. M. Smith (eds.) (Elsevier Science Publishers), pp. 231–248.

Jaynes, E. T. (1958) 'Probability theory in science and engineering', manuscript, (Washington University, Physics Dept.)

Halpern, J. Y. (1986) 'Reasoning about knowledge: An overview', in *Theoretical Aspects of Reasoning about Knowledge*, Proceeding of the 1986 Conference. J. Y. Halpern ed. (Morgan Kaufmann Publishers), pp. 1–18.

Halpern, J. Y. and Y. O. Moses (1985) 'A guide to the modal logics of knowledge and belief', in *Proceedings of the Ninth International Joint Conference on Artificial Intelligence*, pp. 480–490.

Jeffrey, R. (1965) *The Logic of Decision* (New York: McGraw-Hill.) (Second edition 1983.)

Kolmogoroff, A. N. (1933) 'Grundbegriffe der Wahrscheinlichkeit', *Ergebnisse der Mathematik und ihrer Grenzgebiete*, No 2, Berlin.

Lewis, D. (1980) 'A subjectivist guide to objective chance', in *Studies in Inductive Logic and Probability II*, Jeffrey R. C. (ed), University of California Press, pp. 263–293. Reprinted in IFS, W. L. Harper, Stalnaker and G. Pearce (eds.) (Boston; Reidel 1981.)

Miller, D. (1966) 'A paradox of information', *British Journal for the Philosophy of Science* **17**.

Shafer, G. (1976) *A Mathematical Theory of Evidence* (Princeton University Press.)

Skyrms, B. (1980A) 'Higher order degrees of belief', in *Prospects for Pragmatism* (Essays in honor of F. P. Ramsey) D. H. Mellor (ed.) (Cambridge University Press.) pp. 109–137.

Skyrms, B. (1980B) *Causal Necessity* (Appendix 2), (Yale University Press.)

Skyrms, B. (1985) 'Maximum entropy inference as a special case of conditionalization', *Synthese* **63**, pp. 55–74.

van Fraassen, B. (1980) 'A temporal framework for conditionals and chance', *Philosophical Review* **LXXXIX**, pp. 91–108.

van Fraassen, B. (1984) 'Belief and the will', *Journal of Philosophy* **81**, pp. 235–256.

RICHARD JEFFREY

CONDITIONING, KINEMATICS, AND EXCHANGEABILITY

PREVIEW

The change ("conditioning") from prior P to posterior $Q = P(\ |E)$ is appropriate only if it changes no probabilities conditionally on E. Under similar conditions a generalization of conditioning ("probability kinematics") is appropriate when $Q(E) < 1$. That generalization is pretty nearly equivalent to ordinary conditioning on the extraordinary proposition that $Q(E)$ has a certain value. Whether or not generalized conditioning is sensitive to the order in which successive changes are made depends on how the changes are set, e.g., by probabilities, or by ratios of probabilities. In a finitistic framework simple and generalized ("partial") exchangeability are characterized and related to probability kinematics.

CONDITIONING[1]

Sometimes an experiment or observation is adequately represented by partitioning the sample space in such a way as to satisfy both of the following conditions.

(1) CERTAINTY: *observing the outcome drives your probability for some cell E of the partitioning to* 1.
(2) SUFFICIENCY: *probabilities conditioned on E remain unchanged.*

Where P and Q are your prior and posterior probability functions, these conditions say that for some cell E and all hypotheses H in the common domain of P and Q, (1) and (2) hold:

$$(1)\ Q(E) = 1 \qquad (2)\ Q(H|E) = P(H|E)$$

$$(3)\ Q(H) = P(H|E)$$

In any such case (3) holds as well, i.e., Q is obtainable from P by conditioning. In fact,

(1) *and* (2) *are jointly equivalent to* (3), *provided* $P(E) > 0$.

Brian Skyrms and William L. Harper (eds.), Causation, Chance, and Credence, Vol. I, 221–255.

Proof. By (1), $Q(H|E) = Q(H)$; from this and (2), (3) follows. Conversely, (3) implies $Q(E) = P(E|E) = 1$, i.e., (1); and by (3), $Q(H|E) = Q(HE)/Q(E) = P(HE|E)/P(E|E) = P(H|E)$, i.e., (2).

The sufficiency condition (2) can be expressed in other forms, which are sometimes more transparent:

(4) *Odds between propositions that imply E don't change: If G and H imply E then* $Q(H)/Q(G) = P(H)/P(G)$.
(5) $Q(H)/P(H) = Q(E)/P(E)$ *if H implies E.*
(6) $Q(s)/P(s)$ *is constant for all sample points s in E.*

Again, it is understood that denominators, e.g. $P(H)$ and $P(E)$ in (5), don't vanish. Of course, the equivalence of (2) with (6) is asserted only where the sample space is countable and P and Q are defined on all of its subsets (sc., *the discrete case*). Summary:

Conditions (2), (4) *and* (5) *are equivalent to each other, and, in the discrete case, to* (6).

Proof. To get (6) from (5) set $H = \{s\}$. Conversely, if $Q(s)/P(s) = k$ for all s in E as in (6) then $Q(E) = \Sigma_{s\in E}kP(s) = kP(E)$ and if H implies E, $Q(H) = \Sigma_{s\in H}kP(s) = kP(H)$, whence (5) follows. Then (5) and (6) are equivalent in the discrete case. Now verify a circle of implications as follows. (2) $\vdash$ (4): if G and H imply E then where X is G or H, $Q(X|E) = Q(X)/Q(E)$ and $P(X|E) = P(X)/P(E)$ or, by (2), $Q(X)/P(X) = Q(E)/P(E)$, whence, (4). (4) $\vdash$ (5): set $G = E$. (5) $\vdash$ (2): put HE for H in (5) to get $Q(HE)/Q(E) = P(HE)/P(E)$, whence, (2).

DIACHRONIC COHERENCE

Bruno de Finetti's (1937) Dutch book argument for the definition $P(H|E) = P(HE)/P(E)$ of conditional probability answers the question of how your acceptance of simple bets is to connect with your synchronic acceptance of conditional bets. David Lewis[2] answers a different question: When should your new probability distribution, Q, be your old distribution, P, conditioned on E?

Both arguments identify your probability for truth of a hypothesis with the price at which you would be willing to buy or sell a ticket that's worth 1 to 0 units of currency depending on whether the hypothesis is true or false. More generally, where the winning ticket is not worth 1 unit, your probability is the fraction p of its winning worth at which you'd buy or sell the ticket; e.g., see ticket 2 below, where the winning worth is p. Your conditional probability for truth of a hypothesis H on condition D is identified with your buying-or-selling price p for a ticket that's worth 1 unit if DH is true, 0 units if $D-H$ is true, and p units if D is false; e.g., see ticket 1 below.

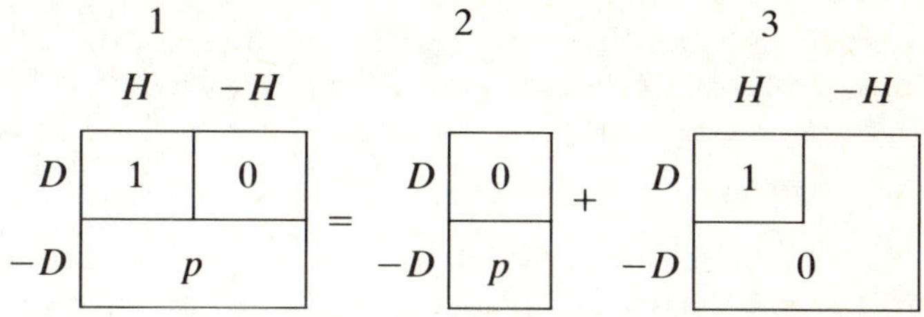

Price: p $\quad pP(-D) \quad P(DH)$

In a version that I borrow from Brian Skyrms (1980b), de Finetti's Dutch book argument is the observation that, come that may, ticket 1 is worth exactly the same as tickets 2 and 3 together; and that therefore, your price for 1 ought to be the sum of your prices for 2 and 3: $p = pP(-D) + P(DH)$. Solving for p we find that $p = P(DH)/P(D)$. Q.E.D.

That argument for the relationship $P(H|D) = P(DH)/P(D)$ between your synchronic conditional and unconditional probabilities is less rough than the characterizations of probabilities in terms of money. The argument depends only on those characterizations sometimes holding good – on the assumption that the laws of probability should be no different when those characterizations fail than when they hold.

Underlying Lewis's diachronic argument is the thought that if you have no other use for the money until tomorrow, there is nothing to choose between buying ticket 1 today, when your probability distribution is P, and buying tickets 2 and 3 tomorrow, when your probability distribution is Q – provided you pay the same for 2 and 3 that you would have paid for 1. If today's and tomorrow's prices correspond to P and Q, you'll pay

$$P(H|D) \qquad P(H|D)Q(-D) \qquad Q(HD)$$

for the respective tickets. The first of these equals the sum of the other two iff $P(H|D) = Q(HD)/Q(D) = Q(H|D)$, i.e., the sufficiency condition (2).

Now where the change between P and Q will be due to your learning the truth value of D tomorrow morning, consider two cases. Case 1: D is false. Then $Q(D) = 0$. The three prices are

$$P(H|D) \qquad P(H|D) \qquad 0$$

– i.e., exactly what the respective tickets are worth. Then case 1 is vacuous; it places no restrictions on P and Q. Case 2: D is true. Then $Q(D) = 1$, i.e., the certainty condition (1). The three prices are

$$P(H|D) \qquad 0 \qquad Q(H)$$

and the condition under which there's nothing to choose between buying ticket 1 today and buying tickets 2 and 3 tomorrow is that $Q(H) = P(H|D)$. Q.E.D.

Note that this conclusion (3) was obtained under assumptions (1) and (2) of certainty and sufficiency. As Teller (1976) points out, the point of the sufficiency requirement is to ensure that in learning that D is true you learn nothing that would change your odds between different ways in which D might come true; see (4).

In the following argument of Lewis's (see Teller 1976) the sufficiency condition plays no explicit role.

Suppose that through an observation or experiment, or in some other way, you expect to learn which cell of a partition is true. You are to announce a complete strategy for changing your probabilities in response to the forthcoming information. Your announcement will specify a new probability distribution for each possibility as to the truth. Suppose that part of what you've announced is that your distribution will be Q if the truth is D.

LEWIS: *If $P(D) \neq 0$ some Dutch strategy will be acceptable to you unless $Q(H) = P(H|D)$.*

Proof. Suppose that $Q(H)$ is too big:

$$Q(H) - P(H|D) = a > 0$$

Then the following three tickets, with their accompanying buying or selling instructions, make up a Dutch strategy. (If, instead, $Q(H)$ is too small, so that a is negative, just reverse the buying/selling instructions.)

1	2	3
Worth \$1 or \$0 or $\$P(H\|D)$ depending on whether DH or $D-H$ or $-D$ is true. Price: $\$P(H\|D)$	Worth $\$a$ if D is true. Price: $\$aP(D)$	Worth \$1 if H is true. Price: $\$Q(H)$
SELL NOW	BUY NOW	BUY IF AND WHEN D PROVES TRUE

To verify Dutchness of the strategy, work out your gains and losses from those transactions in the relevant cases, bearing in mind that $P(H|D) - Q(H) = -a$:

CASE	GAIN FROM 1	GAIN FROM 2	GAIN FROM 3	NET GAIN
DH	$P(H\|D) - 1$	$a-aP(D)$	$1-Q(H)$	$-aP(D)$
$D-H$	$P(H\|D)$	$a-aP(D)$	$-Q(H)$	$-aP(D)$
$-D$	0	$-aP(D)$	0	$-aP(D)$

Lewis offers a strategy for making one or the other of two books, depending on circumstances: your book is {Sell 1, Buy 2, Buy 3} or {Sell 1, Buy 2}, depending on whether D is true or false. And since there is no gain or loss from ticket 1 if D is false, the second book has the same value as the singleton {Buy 2}. The sale of 1 and the acquisition of 2 both seem acceptable today, when your probabilities are given by P; and if you acquire 3 tomorrow that will seem fair then, for you will buy 3 only if D is true, hence, only if its price $\$Q(H)$ is acceptable to you then.

Where does the sufficiency condition enter Lewis's argument? It enters with the assumption that today, while your probability distribution is P, you know what tomorrow's Q will be if you learn that D is true. Then your odds between ways in which D might come true are determined by today's judgments: by P. That's why tomorrow's odds between propositions that imply D are the same as today's. That's where the sufficiency condition enters.

SUFFICIENCY AND PAROCHIALISM

When should P be updated to Q by conditioning? The right answer is the sufficiency condition. A different answer, commonly given, is that you should always update by conditioning, i.e., on the strongest proposition E *in the given sample space* to which you now attribute probability 1. That answer is often right, but not always. It's wrong when the E in question violates the sufficiency condition. As an attractive but unreliable rule of probabilistic inference, it's a fallacy. Call it "parochialism".

Clearly, mere certainty isn't enough, e.g., if you learn that E and F are both true, you'll be certain of E, of F, and of their conjunction EF, and yet the conditional probability functions P_E, P_F and P_{EF} may all be different. If EF is the strongest proposition in the space of which you're now certain, parochialism would have your new probability be P_{EF}. But that may be wrong: your new probability Q may rightly differ from P_{EF}. The following well-known horror story illustrates how such clashes are sometimes reparable by enriching the sample space.

THE THREE PRISONERS. Two are to be shot and the other freed; none is to know his fate until the morning. Prisoner A asks the warder to confide the name of one other than himself who will be shot, explaining that as there must be at least one, the warder won't be giving away anything relevant to A's own case. The warder agrees, and tells him that B will be shot. This cheers A up a little, by making his judgmental probability for being freed rise from 1/3 to 1/2. But that's silly: A knew already that one of the others would be shot, and (as he told the warder) he's no wiser about his own fate for knowing the name of some other victim.

Diagnosis. A goes wrong by using the 3-point sample space $\{a, b, c\} = \{A \text{ lives}, B \text{ lives}, C \text{ lives}\}$. In that space the warder's statement does eliminate one point, b, and doesn't eliminate either of the others. Following the counsel of parochialism, A conditions on $E = \{a, c\}$. With equal prior weights for the three points, A thus shares the missing 1/3 out equally between a and c. To see why that's wrong we need an ampler sample space, in which one can directly represent propositions about what the warder says; for A's evidence isn't just that B will die, but that the truthful warder names B as one who is to die.

A 4-point space will do if we don't insist on a uniform probability distribution. The points are (say) $a\tilde{b}$, $a\tilde{c}$, $b\tilde{c}$, $c\tilde{b}$, where $x\tilde{y}$ identifies X as the prisoner who is to go free and Y as the one the warder names. Antecedently, $b\tilde{c}$ and $c\tilde{b}$ each get weight 1/3, and the weights of $a\tilde{b}$ and $a\tilde{c}$ sum to 1/3: the way in which that 1/3 is parcelled out between them expresses A's opinion about the warder's policy for naming B or C when he can honestly do either.

Presumably A thinks that when the warder can honestly name either B or C as doomed, he's as likely to name one as the other. We'll see that in this case the warder's naming B shouldn't change A's judgment about his own prospects.

TABLE I
Sample space for the three prisoners problem

		Warder names B	Warder names C
	A lives	$a\bar{b}$ (1/6)	$a\bar{c}$ (1/6)
A dies	B lives		$b\bar{c}$ (1/3)
	C lives	$c\bar{b}$ (1/3)	

The numbers in Table I are A's prior probabilities for sample points. A's new probability for living ought to be his old conditional probability given that *the warder names B:*

$$Q(A \text{ lives}) = P(A \text{ lives} \mid \text{Warder names } B) = 1/3$$

That's the same as $P(A \text{ lives})$. A's mistake was to condition on the proposition $E = \{a\bar{b}, c\bar{b}, a\bar{c}\}$ that B will die, i.e., $\{a, c\}$ in the 3-point space, instead of on the stronger proposition $\{a\bar{b}, c\bar{b}\}$ that the warder (truthfully) names B – a proposition that corresponds to no subset of the 3-point space. The trouble with E is that although $Q(E)$ is and ought to be 1 after A hears the warder name B, A's new probabilities $Q(H \mid E)$ conditionally on E won't all agree with the corresponding prior conditional probabilities, $P(H \mid E)$. In either space sufficiency (2) fails for that E – as does the condition of parochialism in the 4-point space. What misleads A is the fact that the condition of parochialism does hold for $E = \{a, c\}$ in the 3-point space.

How do we know that the sufficiency condition fails?

One way is to note that version (4) of that condition fails when E is the proposition $\{a\bar{b}, c\bar{b}, a\bar{c}\}$ that B will die. Hearing the warder name B, not C, A's probability for $a\bar{c}$ drops from $P(a\bar{c}) = 1/6$ to $Q(a\bar{c}) = 0$, while his probability for the warder's naming B climbs from $P(a\bar{b}, c\bar{b}) = 1/2$ to $Q(a\bar{b}, c\bar{b}) = 1$. Then A's odds between propositions $\{a\bar{c}\}$ and $\{a\bar{b}, c\bar{b}\}$ that imply E change from a positive value (1:3) to zero.

Another way is to note that version (5) fails: $Q(H)/P(H)$ vanishes when H is $\{a\bar{c}\}$, but not when H is $\{a\bar{b}, c\bar{b}\}$.

Still another is to note that version (6) fails: $Q(s)/P(s)$ vanishes when s is $a\bar{c}$, but not when s is (say) $a\bar{b}$.

PROBABILITY KINEMATICS

We have seen that natural enlargements of sample spaces can yield propositions satisfying both conditions: sufficiency, and parochialism too. But that doesn't always work.

Probability kinematics is a generalization of conditioning, apt in some cases where observation or experiment or reflection prompts redistribution of probabilities over a countable[3] partition **E** of a sample space without driving the probability of any cell all the way to 1. As in conditioning, we require that the observation (or whatever) change the judgmental probability distribution over a partition without changing any probabilities conditionally on cells E of the partition:

(7) SUFFICIENCY: $Q(H|E) = P(H|E)$ *for each* E *in* $\mathbf{E}$ *for which* $Q(E) > 0$.

Here we suppose that $P(E)$ is positive for each cell E: there is no possibility of raising $P(E)$ from 0 to a positive value $Q(E)$. But certainly $Q(E)$ can be 0 where $P(E)$ is positive. In that case, where $Q(HE)/Q(E)$ is indeterminate, we assign $Q(H|E)$ the value $P(H|E)$ so that condition (2) will be satisfied.

Sufficiency (7) corresponds to (2). Condition (1) is dropped: now the observation need not identify the true cell. Sufficiency is equivalent to the following condition.

(8) KINEMATICS: $\quad Q(H) = \Sigma_E Q(E)P(H|E)$

Proof. As Q is a probability measure, $Q(H) = \Sigma_E Q(E)Q(H|E)$. Then (7) implies (8). For the converse, consider any cell D of **E**. By (8) we have $Q(HD) = Q(D)P(HD|D)$ since $P(HD|E) = 0$ for each cell $E \neq D$, whence $Q(H|D) = P(H|D)$.

The kinematical formula (8) is the promised generalization of conditioning. Ordinary conditioning is the special case of (8) in which $Q(D) = 1$ for some cell D and so $Q(E) = 0$ for all $E \neq D$.

In the discrete case, (7) is equivalent to[4]

(9) $Q(s)/P(s) = Q(E_s)/P(E_s)$, *where* E_s *is the cell of* $\mathbf{E}$ *that* s *lies in.*

Proof. With $H = \{s\}$ and $E = E_s$, (7) immediately yields $Q(s)/Q(E_s) = P(s)/P(E_s)$, i.e., (9). For the converse, note that in the discrete case $Q(HE) = \Sigma_{s \in HE} Q(s)$, so that (9) implies $Q(HE) = \Sigma_{s \in HE} P(s)Q(E_s)/P(E_s) = [Q(E)/P(E)]\Sigma_{s \in HE} P(s) = [Q(E)/P(E)]P(HE)$, whence (7) is immediate.

Our earlier proofs that (2) is equivalent to (4) and to (5) also prove that (7) is equivalent to each of the following.

(10) *For each E in* **E**, *odds between propositions that imply E don't change.*

(11) *For each E in* **E**, *$Q(H)/P(H)$ is constant for all H that imply E, provided $P(H) > 0$.*

In statistical jargon, condition (7) says that the partition **E** is sufficient for the pair $\{P, Q\}$, and that is the same as saying that some statistic T that has the cells of **E** as its sets of constancy is sufficient for $\{P, Q\}$. Where **E** has n cells, the values of T might be the numbers 0 through $n - 1$, or they might be any other n distinct objects, e.g., most simply, the very cells of the partition: $Ts = E$ iff $s \in E \in \mathbf{E}$.

The ratio $Q(s)/P(s)$ in (9) is a useful statistic. Its importance lies in the fact that it is a minimal sufficient statistic for $\{P, Q\}$, in this sense:

MINIMALITY. *Among partitions of a countable space that are sufficient for $\{P, Q\}$ there is a coarsest, i.e., the one whose cells are the sets of constancy of the statistic r, i.e., $r(s) = Q(s)/P(s)$.*

Proof. Let **E** be a partition sufficient for $\{P, Q\}$. By (9), r is constant on each cell of **E**. Thus each of r's sets of constancy is a union of cells of **E**, and the collection **R** of r's sets of contancy is seen to be a partition at least as coarse as any partition that's sufficient for $\{P, Q\}$. That **R** is sufficient for $\{P, Q\}$ follows from the equivalence of (7) to (9) and the fact that (9) is satisfied when **E** = **R**.[5]

CONDITIONING ON FUTURE JUDGMENTS

If the sample space is rich enough, probability kinematics is representable as a form of conditioning, i.e., conditioning on the datum D that your new distribution Q will assign values $Q(E_i) = q_i$ to the cells E_i of a partition **E**.[6]

Here is Brian Skyrms's version of the argument.[7] According to the Reflection Principle (van Fraassen's (1984) term), D will be reflected back into your prior distribution P as in (12) below. Suppose, too, that H is independent of D conditionally on each cell of **E**, as Skyrms's principle (13) stipulates.

(12) REFLECTION: $P(E_i|D) = q_i$ if $P(D) > 0$
(13) SUFFICIENCY: $P(H|E_iD) = P(H|E_i)$ if $P(E_iD) > 0$

Now if $Q(H) = P(H|D)$, the kinematical formula (8) follows from (12) and (13) by straightforward substitution into the law

$$P(H|D) = \Sigma_i P(H|E_iD)P(E_i|D)$$

of the probability calculus. Thus conditioning suffices for the purposes that probability kinematics is meant to serve, if we'll condition on our own future states of judgment conformably with (12) and (13).

As Skyrms points out, condition (13) isn't met for all H, and shouldn't be. In an extreme example H is the hypothesis that $Q(E_i) = q_i$, and it's not D but E_i that's irrelevant to H; for where D implies H, $P(H|E_iD) = 1$ but perhaps $P(H|E_i) < 1$. And the case is similar whenever it's final (Q) probabilities rather than the corresponding unknown facts that are initially (P) seen as relevant to H, as in Skyrms's example:

> my current probability that I will sweat at the moment of arriving at my final probability, conditionally on the fact that Black Bart will not really try to gun me down *and* that that my final probability that he will try to kill me will be 0.999, is *not* equal to my current probability that I will sweat, conditional on the fact that he will not really try to gun me down. The sweating is highly correlated with my final degree of belief rather than the fact of the matter.[8]

But normally there is a natural Boolean algebra **A** of "objective" hypotheses H for which sufficiency (13) holds.

Diaconis and Zabell (1982) observe that when Q comes from P by conditioning on D, $1/P(D)$ is an upper bound on the ratios of your new to old probabilities on **A**: $Q(H)/P(H) \leqslant 1/P(D)$. That's so because $Q(H) = P(H|D) = P(HD)/P(D)$, so $Q(H)/P(H) = P(D|H)/P(D)$, where $P(D|H) \leqslant 1$. They prove that this condition is necessary and sufficient for Q to be obtainable by extending P beyond **A** and then cutting back to **A** by conditioning on a proposition outside **A**. I'll call that process "superconditioning".

SUPERCONDITIONING

It will make for clarity if we distinguish three Boolean algebras: the algebra **A** of objective hypotheses on which today's and tomorrow's judgmental probability distributions P and Q are defined; the algebra **D** of subjective hypotheses (notably, the hypothesis D about Q) about which you are unlikely to have any judgmental probabilities; and an overarching algebra **X** in which the other two are somehow embedded. The only information we'll need about today's probabilities on **D** is the probability – p, let's say – of D itself. When we condition on D to get the effect of probability kinematics, the exact value of p is of no importance, as long as it is not 0 or 1: not 0, so that we can condition on D, and not 1, so that conditioning on D can make a difference. Instead of "P" we'll use a different designation, "M", for the probability function on **X** that sums up P and p.

If D is of interest only as a bridge over which to pass from P to Q, both of which are defined on **A**, the question arises: under what conditions on P and Q can **A** be mapped into some algebra **X** on which a probability measure M can be cooked up out of P and Q, which, conditioned on some member X of **X**$-$**A**, yields on the image of **A** in **X** a distribution that's the image of Q? Answer:

(14) DIACONIS AND ZABELL: *P yields Q by superconditioning (defined below) iff there is an upper bound $b \geq 1$ on the ratios $Q(H)/P(H)$ for propositions H in* **A**.

Definition. (All algebras are Boolean σ-algebras.) P yields Q by superconditioning iff there exist: a Boolean σ-isomomorphism f from the algebra **A** on which P and Q are defined onto a subalgebra of an algebra **X**; a probability measure M on **X** that corresponds to P on **A**; and an X in **X** that gets Q out of M by conditioning: $M(f(H)) = P(H)$, $M(X) > 0$, $Q(H) = M(f(H)|X)$.

Proof of (14). (Throughout, "H" ranges over **A**.) If Q comes from P by superconditioning on X in **X** then

$$Q(H) = M(f(H)|X) \leq M(f(H))/M(X) = P(H)/M(X),$$

i.e., $Q(H)/P(H) \leq b$ with $b = 1/M(X) \geq 1$. For the converse, suppose b satisfies the conditions, with **A** an algebra (i.e., a Boolean σ-algebra) of subsets of W. For $x \neq y$ not in W, define

$$f(H) = H \times \{x, y\} \qquad X = W \times \{x\}$$
$$\mathbf{D} = \{f(W), X, f(W) - X, \emptyset\}$$

$\mathbf{X}$ = all countable unions of sets of forms $f(H) \cap X$, $f(H) - X$. So defined, $\mathbf{X}$ is an algebra of subsets of $f(W)$, and f respects complementation and union: $f(W - H) = f(W) - f(H)$, $f(\cup_i H_i) = \cup_i f(H_i)$. The idea is that each point $\langle w, x\rangle$ or $\langle w, y\rangle$ of $f(W)$ tells the same story that w does, with the extra information that X is true (x) or false (y). Then $f(H) = \{\langle w, z\rangle : w \in H, z = x, y\}$ gives the same information that H does, i.e., $(f(H) \cap X) \cup (f(H) - X)$. Now define

$$M(f(H) \cap X) = Q(H)/b,$$
$$M(f(H) - X) = P(H) - Q(H)/b.$$

The conditions on b ensure that these equations determine M as a probability measure on X. The three conditions at the end of the definition of superconditioning then hold.

QUALMS ABOUT REFLECTION

Goldstein (1983, 1985) and van Fraassen (1984) provide Dutch book arguments for reflection principles. The following argument for (12) is a minor variant of van Fraassen's.

VAN FRAASSEN: *if* $P(D) \neq 0$, *where* D *specifies new values* $Q(E_i) = q_i$ *for the cells of a partition* $\{E_i : i = 1, \ldots, n\}$ *then a Dutch program will be acceptable to you unless* $P(E_i|D) = q_i$.

Proof. Set $H = E_i$ and $Q(H) = q_i$ in Lewis's book with $a = q_i - P(E_i|D) > 0$:

1	2	3
Worth \$1 or \$0 or \$$P(E_i\|D)$ depending on whether DE_i or $D - E_i$ or $-D$ is true. Price: \$$P(E_i|D)$	Worth \$$a$ if D is true. Price: \$$aP(D)$	Worth \$1 if E_i is true. Price: \$$q_i$
SELL NOW	BUY NOW	BUY IF AND WHEN D PROVES TRUE

If D is false (Lewis's third case) bets 1 and 3 are off, and you lose \$$aP(D)$ on 2. If D is true (Lewis's first two cases) you lose \$$aP(D)$ in one way or another. Conclusion: the book is Dutch.

But Van Fraassen (1984) calls attention to a troublesome consequence of the reflection principle. To highlight the problem, consider Brian Skyrms' (1987) example:

THE WRATH OF KHAN. If I am now convinced that tomorrow the mindworm of the treacherous Khan will change my probability for H to a certain unreasonable value a, the reflection principle requires me to adopt that unreasonable value already today.

Proof. In (12), set $E_1 = H$. To this D assigns a as its new, Q probability. But by (12) a can't be new after all if $P(D) = 1$, for then $P(H|D) = P(H) = q_1 = a$.

The problem isn't that you are conditioning on D. In the present case, as in the problem of the three prisoners, there's a stronger proposition that you're certain of, i.e., here, the Mindworm hypothesis M that both implies and explains D's truth. Although the sufficiency condition fails in the form $Q(\ |D) = P(\ |D)$ perhaps it holds with M in place of D. Perhaps; but that's no help. Even if you condition on M, $P(D)$ will be 1, and that's enough to make the reflection principle imply that $P(E_i) = q_i$.

As I understand him, van Fraassen holds that in the light of such examples the reflection principle is to be rejected in case your future judgmental state is thus alien to you. What goes wrong in the Wrath of Khan example is the fact that today's and tomorrow's judgmental states are different in the way in which two different people's states might be different. There is a discontinuity between your points of view today and tomorrow that isn't accounted for by the different values that P and Q assign, but by your present view of the source of that difference. It's not that you can think of no reasonable way in which a might come to be your probability for H tomorrow, but that in fact you think the change will not come about in any such way: the hypothesis R that the transition from P to Q is a reasonable one rules D out.

(15) REFLECTION RESTRICTED. $P(E_i|R) = q_i$ *if* $P(D|C) > 0$ *with D as above and with R as the hypothesis that your P-to-Q transition is reasonable.*

But what about the less extreme cases, where you are less than certain of D's truth, or where the Mindworm story is replaced by some more homely hypothesis about fear, fatigue, drugs, alcohol, etc.? Do you get into trouble by merely entertaining one of these hypotheses, i.e., by having $P(D) \neq 0$ when you regard D as pathological?

No. To entertain an analog of the Mindworm hypothesis is just to give some credence to the possibility that you really will be in trouble tomorrow – real trouble, that you don't get into just by entertaining the hypothesis that it may come about. To see that more clearly, let's review the argument for (13), to see where it goes wrong when $P(D|R) = 0 < P(D)$. We'll see that where $P(D|R) = 0$ your expectation of gain from books that would have 0 expectation under happier circumstances will be negative, and a Dutch book may well have higher expected utility than any available non-Dutch book. Too bad, if so. But it's not thinking that makes it so. Here are the details.

Today, when your probability function is P, the Dutch strategy won't be yours. Today you'll sell 1 and buy 2, but you won't instruct your broker to buy 3 if and when D proves true. You know better. Of course, if you are acting as your own broker tomorrow, and D proves true, you will then regard 3 as a bargain. And you'll then view today's judgment as an aberration from which you have fortunately recovered; you'll buy 3, knowing that you'll thereby suffer the same overall loss of $\$aP(D)$ that you would have suffered if D had been false. As you see already, that will look sensible to you because your expectation if you do *not* buy 3 would then be the same, i.e., $P(E_i|D) - Q(H)$ from the sale of 1, added to a known gain of $a - aP(D)$ from buying 2, a winner.

That's sad; and you're saddened by it already, today, for today you think that if D is true you'll be gulled tomorrow; and you even see today that whether D is true or false, tomorrow you'll think you were gulled today, when you allowed yourself to be guided by the function P in deciding to sell 1 and buy 2 – transactions from which your (P) expectation was 0 but from which you now see that tomorrow's expectation will be $-aP(D)$ on either hypothesis about D.

That's sad, but that's life. If your judgments today and tomorrow are so alien that on each day you regard yourself as irrational on the other, you are in genuine difficulties. Chemotherapy or psychotherapy might help, but on the mindworm hypothesis and its homely analogs, the laws of probability can't. They can't tell you which (if either) of P, Q is reasonable; nor should you expect them to.

The key is your readiness to identify yourself today with yourself tomorrow. It is only if you do so that the diachronic Dutch strategy in the proof of (12) will be yours, simply. If you don't, today's "you" accepts transactions 1 and 2 but rejects 3 and tomorrow's accepts 3 but reject the first two; and there's no one agent that you both recognize as accepting the whole book.

SUCCESSIVE UPDATING

What happens when the kinematical formula (8) is applied to two different partitions, $\mathbf{E} = \{E_i : i = 1, \ldots, m\}$, $\mathbf{F} = \{F_j : j = 1, \ldots, n\}$, in succession?

$$P \xrightarrow{\mathbf{E}} Q \xrightarrow{\mathbf{F}} R$$

It's straightforward to get from P to R in two steps. First get from P to Q via (8) – which, following Hartry Field (1978)[9] we'll find it useful to rewrite as

$$(17) \qquad Q(H) = P(H)\Sigma_i e_i P(E_i|H), \qquad e_i = Q(E_i)/P(E_i)$$

Then get from Q to R by applying the same transformation again, but with $F_j \in \mathbf{F}$ in place of $E_i \in \mathbf{E}$, and with Q and R in place of P and Q:

$$(18) \qquad R(H) = Q(H)\Sigma_j f_j Q(F_j|H), \qquad f_j = R(F_j)/Q(F_j)$$

To combine these into a single step from P to R, put $Q(F_jH)/Q(H)$ for $Q(F_j|H)$ in (18), and then apply (17) with F_jH in place of H:

$$\begin{aligned} R(H) &= \Sigma_j f_j Q(F_jH) \\ &= \Sigma_j[f_j P(F_jH)\Sigma_i e_i P(E_i|F_jH)] \\ &= \Sigma_j[f_j \Sigma_i e_i P(E_iF_jH)] \end{aligned}$$

or, finally,

$$(19) \qquad R(H) = P(H)\Sigma_{i,j} e_i f_j P(E_iF_j|H), \qquad e_i = Q(E_i)/P(E_i),$$
$$f_j = R(F_j)/Q(F_j)$$

This suggests that the effect of the two steps above is the same as that of one big step from P to R via the product $\mathbf{E} \times \mathbf{F} = \{EF : E \in \mathbf{E}, F \in \mathbf{F}\}$ of the two partitions, where in the big step the weight $g_{ij} = R(E_iF_j)/P(E_iF_j)$ of each cell E_iF_j is the product of the weights that its factors have in the two-step transition:

$$(20) \qquad R(H) = P(H)\Sigma_{ij} g_{ij} P(E_iF_j|H),$$
$$g_{ij} = R(E_iF_j)/P(E_iF_j) = e_i f_j$$

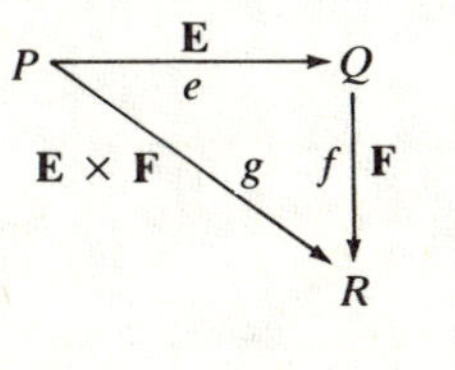

The suggestion is right:

> If **E** is sufficient for $\{P, Q\}$ and **F** is sufficient for $\{Q, R\}$ then $\mathbf{E} \times \mathbf{F}$ is sufficient for $\{P, R\}$; and $g_{ij} = e_i f_j$.

Proof. By definition, $g_{ij} = R(E_i|F_j)R(F_j)/P(F_j|E_i)P(E_i)$. Now under the two hypotheses, $R(E_i|F_j) = Q(E_i|F_j) = Q(E_iF_j)/Q(F_j)$ and $P(F_j|E_i) = Q(F_j|E_i) = Q(E_iF_j)/Q(E_i)$, whence $g_{ij} = Q(E_i)R(F_j)/Q(F_j)$ $P(E_i) = e_i f_j$. Substituting g_{ij} for $e_i f_j$ in (19) we then have (20), according to which $\mathbf{E} \times \mathbf{F}$ is sufficient for $\{P, R\}$.

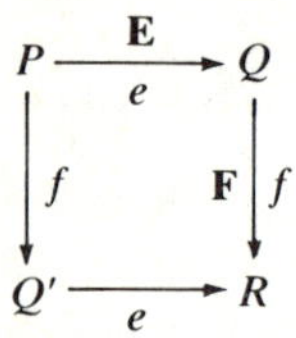

Updating is always commutative when taking a step is a matter of *setting ratios* e_i or f_j of new to old cell probabilities E_i or F_i as in the diagram. Start from P. If we first go down to Q' via **F** and then go right via **E**, we reach the same destination R that we'd have reached by first going right to Q via **E** and then down to R via **F**. That's clear since formula (19) is invariant under permutation of the two steps.

Example. To each point w in the sample space $\{1, 2, 3, 4\}$ the initial distribution P assigns the value $w/10$, as in the upper left of Table II. The partitions are $E_1 = \{1, 2\}$, $E_2 = \{3, 4\}$ (rows) and $F_1 =$

TABLE 2
Changes are commutative when specified by ratios

P	1/2	1/2		Q	19/42	23/42
3/10	1/10	2/10	$e_1 = 5/3$	1/2	1/6	1/3
7/10	4/10	3/10	$e_2 = 5/7$	1/2	2/7	3/14
	$f_1 = 21/19$	$f_2 = 36/23$			$f_1 = 21/19$	$f_2 = 36/23$
Q'	21/38	17/38		R	1/2	1/2
11/38	21/190	34/190	$e_1 = 5/3$	655/874	7/38	13/23
27/38	84/190	51/190	$e_2 = 5/7$	219/874	6/19	54/161

$\{1, 4\}$, $F_2 = \{2, 3\}$ (columns). Cell probabilities are shown in the margins, e.g. $P(E_2) = 7/10$. Suppose we go right first, making the **E** cells equiprobable, and then down, making the **F** cells equiprobable. That's a trip from P to Q to R, described in terms of the new cell probabilities: $Q(E_i) = 1/2$, $R(F_j) = 1/2$.

Let's now redescribe the two changes in terms of ratios. The business of making the **E** cells equiprobable is a matter of setting $e_1 = 5/3$, $e_2 = 5/7$; and thereafter, making the **F** cells equiprobable amounts to setting $f_1 = 21/19$, $f_2 = 36/23$. If instead we had gone from P to Q' by setting the f's just as we did in the second change above we'd have $21/19 = f_1 = Q'(F_1)/P(F_1) = 2Q'(F_1)$, so that $Q'(F_1) = 21/38$ and $Q'(G_2) = 17/38$. We then get from Q' to R if we set $e_1 = 5/3$ and $e_2 = 5/7$ as before. We obtain commutativity by setting ratios of new to old probabilities in the same way, regardless of order. To do so is in effect to set new probabilities – but to set them at values that depend on the order in which the changes are made.

Updating is not generally commutative when each step is a matter of *setting the new probabilities* q_i or r_j of the cells of a partition. The reason is that when we update by first changing probabilities of the **F** cells to r_j and then changing probabilities of the **E** cells to q_i, formulas (17), (18) and (19) give way to

$$
\begin{aligned}
S(H) &= P(H)\Sigma_j[r_j/P(F_j)]P(F_j|H)\\
T(H) &= S(H)\Sigma_i[q_i/S(E_i)]S(E_i|H)\\
&= P(H)\Sigma_{ij}[q_ir_j/S(E_i)P(F_j)]P(E_iF_j|H)
\end{aligned}
$$

Comparing the one-step version of $T(H)$ with (19) we see that while the partition is the same, i.e., $\mathbf{E} \times \mathbf{F}$, the parameter $e_if_j = q_ir_j/P(E_i)Q(F_j)$ has been replaced by a parameter that will be different except when *for all* i, j, $S(E_i)/P(E_i) = Q(F_j)/P(F_j)$.[10]

FROM EPISTEMOLOGY TO STATISTICS

Probability kinematics was first introduced for an in-house philosophical purpose: to show how, in principle, all knowledge might be merely probable, in the face of *a priori* arguments to the contrary, e.g., those of C. I. Lewis (1946, p. 186), who saw conditioning as the only reasonable way to modify judgmental probabilities by experience. According to Lewis:

The data which themselves support a genuine probability, must themselves be certainties. We do have such absolute certainties, in the sense data initiating belief and in those passages of experience which later may confirm it. But neither such initial data nor such later verifying passages of experience can be phrased in the language of objective statement – because what can be so phrased is never more than probable.

Using probability kinematics, I aimed to show (1957, 1965) how the familiar language of objective statement needed no supplementation by what C. I. Lewis (*loc. cit.*) called "the expressive use of language, in which what is signified is a content of experience and what is asserted is the givenness of that content." I saw myself as defending what Carnap had called "physicalism" against phenomenalistic strictures – including, as it turned out, strictures that Carnap himself was to urge against my use of kinematics.[11] The line was what I have lately (1985, p. 114) been calling "radical probabilism", i.e., the epistemological view (essentially, de Finetti's and Ramsey's) that sees judgmental probabilities as our immediate responses to experience, and sees *Protokollsätze* to which we might attribute probability 1 as too thin on the ground to provide a foundation for probable knowledge.

Although it rejects, as misguided, demands for a uniform, nonprobabilistic basis for probability judgments such as relative frequencies are sometimes thought to provide, radical probabilism does undertake to show when and how data about frequencies are reflected in probability judgments – answering the questions "When?" and "How" in terms of structural features of judgmental probability distributions, whose owners would see statistics as relevant to their opinions. This is the point of de Finetti's work on exchangeability and partial exchangeability over the past 50 years and more – work recently pushed further and generalized by Persi Diaconis and David Freedman in a way that connects it closely to probability kinematics. The remainder of this paper is an elementary introduction to that work, in terms of probability distributions over finitary sample spaces.

A REALISTIC SAMPLE SPACE

The usual sample space for coin-tossing masks the absurdity of von Mises' (1919, 1957) idea that the actual tosses of a coin are simply a readily accessible finite initial segment of an infinite sequence of possible tosses – an infinite sequence that tells how all the untried tosses would have come out, had they been tried. The sample space is the set of all omega-sequences of 0's (tails) and 1's (heads). The proposition that the first toss is a head is then represented by the set of all infinite binary sequences that start with 1's; and so on.

What would be a realistic sample space for coin-tossing?

Where a fixed finite number n of tosses is contemplated, the sample space $\{0, 1\}^n$ consisting of the 2^n distinct sequences of n 0's and 1's is adequate and realistic; but where n is unknown, we need to paste together the infinity of such finite spaces to get a realistic substitute $W = \cup_n \{0, 1\}^n$ for the unrealistic, familiar "Mises" space.

The usual binomial probability distributions B_{np} over the finite sample spaces $\{0, 1\}^n$ (for coin-tossing with probability p for heads) can be pasted together in infinitely many different ways to get distributions over W, i.e., for any non-negative a's that sum to 1,

$$P(H) = \Sigma_n a_n B_{np}(H)$$

B_{np} is understood to have its usual values on $\{0, 1\}^n$ and to vanish elsewhere in W; the weight a_n is your probability $P(\{0, 1\}^n)$ for the proposition that there are n tosses in all – a proposition that has no counterpart in the Mises space.

Here's a way of looking at it. Each sample space (for coin-tossing, die-rolling, etc.) is the set W of all "words" $w, x, \ldots$, i.e., finite sequences of "letters" from a finite "alphabet" L. If a, b, c, d, e are in L the words $w = \langle a, b, c\rangle$ and $x = \langle d, e\rangle$ will be written as "*abc*" and "*de*", and their concatenation $w + x = \langle a, b, c, d, e\rangle$ will be written as "*abcde*".

$A + B$ will be the set of all results of adding a word from B on to the end of a word from A:

$$A + B = \{x + y\colon x \in A \text{ and } y \in B\}$$

There is a 0-letter "null" word o, which functions as an identity element:

$$\mathrm{o} + w = w + \mathrm{o} = w.$$

Powers of sets A of words are defined:

$$A^0 = \{\mathrm{o}\}, A^{n+1} = A + A^n.$$

Finally, the realistic sample space W is the result of applying to the alphabet L the operation $*$: where $n = 0, 1, 2, \ldots,$

$$A^* = \cup_n A^n, \qquad W = L^*$$

Example: die-tossing. L is the six-letter alphabet $\{1, 2, 3, 4, 5, 6\}$, representing the possible outcomes of single rolls of a die; L^* represents all possible (finite!) sequences of such outcomes, and subsets of L^* represent various propositions. Thus, $\{1\}^*$ says that if there have been or will be any tosses, they'll all yield aces; $\{2, 3, 4, 5, 6\}^*$ says that the ace never has shown up and never will; and L^n says that the die will have been rolled n times in all. In particular, L^0, i.e., $\{o\}$, is the proposition that the die is never rolled, and L itself is the proposition that the die is rolled just once. $W + \{1\} + W$ says that there is at least one ace (past, present, or future).

In this notation words are interpreted as complete scenarios, covering the past, present, and future, and *nothing marks the present.*[12] Nor does anything mark the present in the more familiar, bizarre scheme in which the actual, finite sequence of outcomes is thought of as an initial segment of an endless sequence, the rest of which shows how untried tosses would have turned out. Note that in the familiar notation there is no way to say how many tosses there are, e.g., our $L + L^*$ (at least 1), or how the last toss turns out, e.g. our $L^* + \{1\}$.

EXCHANGEABILITY[13]

Use the new notation with a finite alphabet L of h letters, which we think of as being given in a definite order. The *tally* of a word w will be a vector

$$(21) \qquad Tw = (N_1w, \ldots, N_hw)$$

where N_iw is the number of times the i-th letter of the alphabet occurs in w. The tally function T is thus a vector-valued statistic. Definition: a probability function P makes the letters of the alphabet *exchangeable* iff words with the same tally are always assigned the same probability:

$$(22) \qquad \textit{Uniformity} \qquad P\{x\} = P\{y\} \quad \text{if} \quad Tx = Ty$$

Calling P "symmetrical" (Carnap 1950, 1962) is another way of saying that P makes all letters exchangeable.

Of course the detailed form of the tally function doesn't matter; all that matters is the corresponding partition of W, i.e., the function's sets of constancy. The key point is that two words belong to the same cell iff one can be transformed into the other by some rearrangement (permutation) of its letters:

(23) $Tx = Ty$ iff y is an anagram of x

If letters are movable, as in scrabble, the cell to which a word y belongs is determined by what you get by dumping y's letters into an urn. Exhaustively drawing without replacement from that urn, you get the successive letters of some word in that cell; and each word in the cell is obtainable in that way.

The possible values of T form the set $\{Tw\colon w \text{ in } L^*\}$. To any such value, say t, the distribution P assigns probability $P(T = t) = P(\{w\colon Tw = t\}) = P(T^{-1}t)$. The equivalence of (7) with (9) in the discrete case implies:

(24) *T is sufficient for any set of symmetrical distributions that assign positive probabilities to the same values of T.*

Note:

(25) *Symmetrical distributions that assign positive probabilities to a value of T become identical when conditioned on that value.*

For if P is symmetrical with tally function T, and if $Tw = t$, then by (22), uniformity,

$$P(w \mid T = t) = 1/(\text{the number of words in } T^{-1}t),$$

which is independent of P.

In particular, for $t = (n_1, \ldots, n_h)$ and $n = \Sigma n_i =$ the length of words with tally t, conditioning on $T = t$ reduces any symmetrical distribution P to the *hypergeometric* probability distribution H_t on L^n that characterizes drawing without replacement from an urn which initially contains n balls, of which n_i ($i = 1, \ldots, h$) are labelled with the i-th letter of L:

(26)
$$\begin{aligned} P(A \mid T = t) &= H_t(A) \\ &= (\text{No. of words in } A \cap T^{-1}t)/(\text{No. in } T^{-1}t) \end{aligned}$$

Furthermore, conditioning on L^n reduces any such P to a definite mixture of hypergeometric distributions H_t (Diaconis 1977):

(27) FINITE DE FINETTI REPRESENTATION

$$P(A \mid L_n) = \Sigma_t a_t H_t(A), \qquad a_t = P(T = t \mid L^n)$$

Example: Tossing a coin twice (Diaconis 1977)

$L = \{0, 1\}$ and t = (number of 0's, number of 1's) where 0's are tails and 1's are heads. Any probability distribution over L^2, symmetrical or not, can be identified by the values it assigns to the sample points 00, 01, 10, 11. The symmetrical ones are those for which the middle two values agree: $P\{01\} = P\{10\}$, in which case $P(\text{Head first}) = P\{10\} + P\{11\} = P\{01\} + P\{11\} = P(\text{Head second})$, $=p$, say. In terms of p, the *binomial* distributions are those symmetrical ones in which $P\{00\} = (1 - p)^2$, $P\{01\} = P\{10\} = p(1 - p)$, $P\{11\} = p^2$. The hypergeometrical distributions form a 3-membered subclass of the symmetrical ones: H_t with t = (2, 0), (1, 1), (0, 2). These correspond to drawing twice without replacement from urns with the compositions shown in Figure 1.

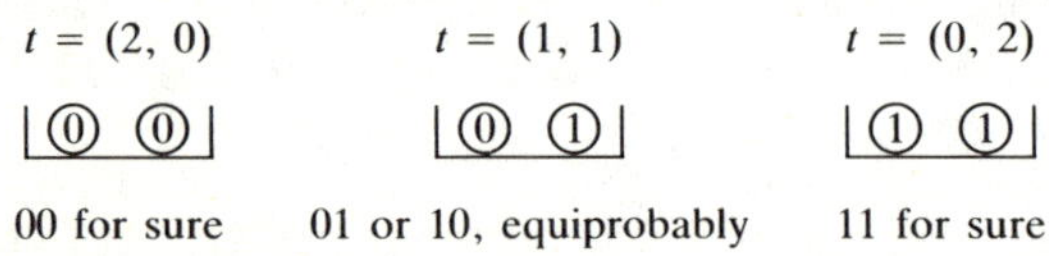

Fig. 1.

The tetrahedron of Figure 2 represents all the probability distributions over $L^2 = \{00, 01, 10, 11\}$. The shaded triangle represents the symmetrical distributions; its vertices, the hypergeometrical ones; the inscribed parabola, the binomial ones. Each point in the (rigid, weightless) tetrahedron would be the center of mass of a unique system of 4

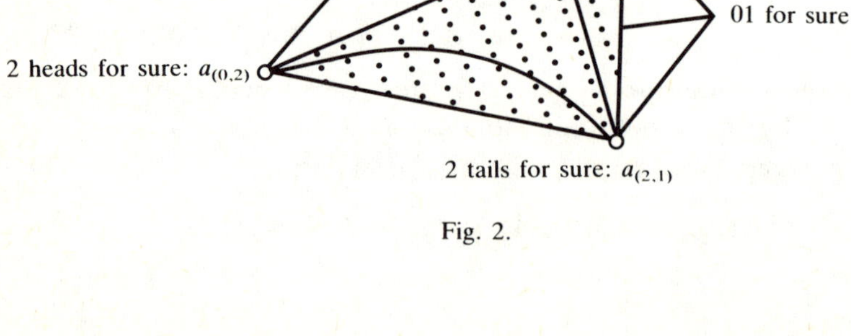

Fig. 2.

masses (summing to 1) that might be fixed to its vertices. In these terms theorem (27) says that each distribution $P(\ |L^2)$ in the shaded triangle is the center of mass of some system of masses a_t fixed to its vertices – the masses $a_t = P(T = t|L^2)$ being determined uniquely by P. (Thus, the Bayes–Laplace–Johnson–Carnap distribution, $P(00) = P(11) = 1/3$, $P(01) = P(10) = 1/6$, determines $a_t = 1/3$ for all 3 t's. That's Carnap's (1950, 1962) m^* with 2 individuals and 1 primitive predicate.)

In addition to its unique representation as a mixture of the 3 hypergeometric distributions that are its vertices, each point inside the triangle and under the parabola is obtainable as a mixture of binomial distributions, in infinitely many different ways. That's because each such point lies on various line segments between points on the parabola, and thus is a mixture of each such pair of (binomial) points. Here is a nongeometrical characterization of the symmetrical distributions on L^2 that are mixtures of binomial distributions.

SUPPES AND ZANOTTI (1980): *Points in the triangle below, above, and on the parabola represent distributions according to which outcomes of the two trials are positively relevant, negatively relevant, and irrelevant to each other.*

It's $P(\text{Two 1's}|L^2) - P(\text{1 on first toss}|L^2)P(\text{1 on second toss}|L^2)$ that determines relevance as positive, negative, or absent.

COROLLARY: *The mixtures of binomials are the P's that don't make the trials negatively relevant to each other.*

For 3 tosses of a coin the 8-point sample space L^3 can be represented by a 7-dimensional simplex, in which the exchangeable distributions form a tetrahedron instead of a triangle – through which the binomial points twist in a curve more complicated than the parabola of Figure 2 (Diaconis 1977).

De Finetti's representation theorem for arbitrary numbers of tosses is most often stated in terms of the unrealistic space L^∞ of omega-sequences of letters:

DE FINETTI'S REPRESENTATION THEOREM. *Every symmetrical distribution on L^∞ is uniquely representable as a mixture of multinomial distributions on L^∞.*

In L^∞, talk about a finite sequence w of outcomes is replaced by talk about the set of all endless prolongations of w. To make that clear we represent the set of all endless prolongations of w by $\{w\} + L^{\infty}$, just as we represent the set of all finite prolongations of w by $\{w\} + L^*$. Thus in (28) and (29) below, $P(\{w\} + L^{\infty}) = P$(the infinite word begins with w) $= P(X_1 = w_1, \ldots, X_n = w_n)$ in the usual notation. For vividness we state the theorem in two special cases: $L = \{0, 1\}$ and $L = \{0, 1, 2\}$.

(28) DE FINETTI'S THEOREM, INFINITE CASE, $L = \{0, 1, 2\}$. *For each symmetrical distribution P on L^{∞} there is a unique probability distribution M over the simplex $S = \{(x, y, z): x + y + z = 1, x, y, z \geqslant 0\}$ of Figure 3 such that for any word w in L^* with tally $Tw = (a, b, c)$,*

$$P(\{w\} + L^{\infty}) = \int_S x^a y^b z^c \,\mathrm{d}M$$

In Figure 3, S is the equilateral triangle with vertices 1 unit from the origin along the three axes. Where $L = \{0, 1\}$ the corresponding simplex is the line segment with ends 1 unit from the origin along the two axes of Figure 4.

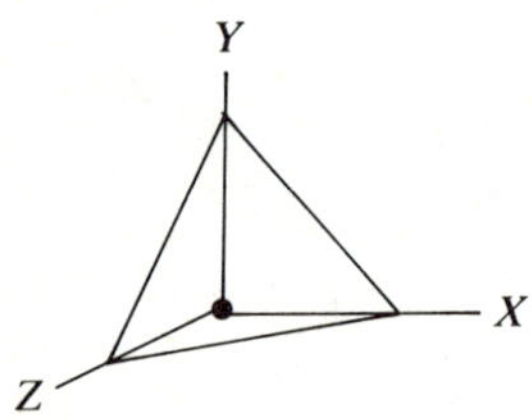

Fig. 3

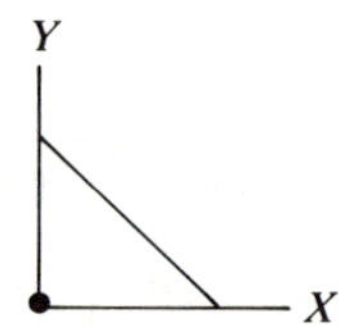

Fig. 4

(29) DE FINETTI'S THEOREM, INFINITE CASE, $L = \{0, 1\}$. *For each symmetrical P on L^{∞} there is a unique probability distribution M over the simplex $S = \{(x, y): x + y = 1, x, y \geqslant 0\}$ of Figure 4 such that for any word w in L^* with tally $Tw = (a, b)$,*

$$P(\{w\} + L^{\infty}) = \int_S x^a y^b \,\mathrm{d}M$$

For the case of a 3-letter alphabet, (28) says that any symmetrical P on L^{∞} assigns to $\{w\} + L^{\infty}$ an M-weighted average of the values $x^a y^b z^c$ that would be assigned to it by the i.i.d. distributions for which the probabilities of the three possible outcomes of a single trial are x, y, z. Thus, any symmetrical P is realizable as the distribution for endless drawings with replacement from an urn containing balls labelled with letters of the alphabet, where the ratio $x:y:z$ of numbers of balls with the different labels was determined randomly in accordance with the distribution M.

For symmetrical P the probability $P(\{w\} + L^\infty)$ as in (28) or (29) serves as an approximation to the corresponding probability in the realistic sample space L^*:[14]

(31) FINITE FORM OF DE FINETTI'S THEOREM.[15]
If P is symmetrical, if the length m of w is less than k, and if $P(\{x + x\}) = 0$ for no word $w + x$ of length k, then (28) *or* (29) *approximates $P(\{w\} + L^* | L^k)$ with error less than $2m/k$ times the size of the alphabet.*

The exact value of $P(\{w\} + L^* | L^k)$ is an M-mixture of the values assigned to $\{w\} + L^{k-m}$ by the hypergeometric distributions H_t where in the 3-letter case $t = (x, y, z)$ with $x + y + z = k$. There H_t are the distributions appropriate for drawing without replacement from urns initially containing k balls marked with the respective letters of the alphabet in the ratios $x:y:z$. $P(\ |L^k)$ is realizable as the distribution for drawing without replacement from such an urn, for which the ratio $x:y:z$ was thought to be determined randomly in accordance with the distribution M. If $w + x$ is much longer than w (if $k \gg m$) then drawing without replacement won't be much different from drawing with replacement (Freedman [1977]), and the expression (28) for $P(\{w\} + L^\infty)$ will closely approximate $P(\{w\} + L^* | L^k)$.[16]

MARKOV EXCHANGEABILITY

The foregoing treatment of exchangeability can be generalized by keeping (22) but dropping (21) – allowing T to be other statistics on L^* that have finite-dimensional vectors as values, provided $Tx = Ty$ is a *congruence relation*, i.e., provided

(31) $\quad T(x + w) = T(y + w) \quad \text{if} \quad Tx = Ty$

for all w, x, y in L^*.[17] Assumption (31) is automatically satisfied in the special case (22) of complete exchangeability. Here we consider a more general case of that sort studied by de Finetti (1938, 1972),

Freedman (1962), Diaconis and Freedman (1980a, b) and Zaman (1984).

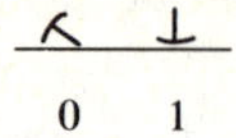

Fig. 5.

Example (Diaconis and Freedman 1980, Diaconis and Zabell 1986): *Tack-Flicking*. A thumbtack is flicked repeatedly – always from the position it landed in after the previous trial, i.e., point down (0) or up (1): see Figure 5. Here a 2-letter alphabet $L = \{0, 1\}$ is apt, with w in $L^* - \{o\}$ reporting the initial position of the tack (first letter) and the outcome of the n-th trial ($n + 1$'st letter). If you think the first outcome may depend on whether the tack starts with point to the floor or not, and think that each outcome is influenced by earlier ones only through its immediate predecessor, your tally function might assign to w in L^* a vector with entries indicating its first letter (i) and the numbers of transitions from down to down (00), down to up (01), etc.:

MARKOV TALLY FUNCTION

$$Tw = t = (i; \#00, \#01, \#10, \#11)$$

Thus $Tw = (0; 1, 2, 1, 0)$ for $w = 00101$ or $w = 01001$, and $Tw = (1; 0, 1, 2, 1)$ for $w = 11010$ or $w = 10110$. In general, where $Tw = (i; a, b, c, d)$, the length of w will be $a + b + c + d + 1$. Then words of the same tally must have the same length. One can show that if $Tx = Ty$ then x and y must end with the same letter. It follows that Markov tally functions T satisfy the congruence condition, (30).

Fig. 6.

For Markov tally functions T over $\{0, 1\}^*$, Diaconis and Freedman (1980, pp. 239–240) contrive urn models that make the words in $T^{-1}t$ equiprobable, as follows. Balls labelled 0 or 1 are drawn without replacement from two urns, themselves labelled 0 and 1. Where i is the first component of t, the first draw is from urn i. Each later draw is from the urn bearing the same label as the ball just previously drawn.

The initial contents of the two urns are determined by t as in Figure 6. You are to keep drawing until instructed to draw from an empty urn. If the other urn is empty, too, *the letter i, followed by the labels on the successive balls, spells out a word in $T^{-1}t$.* That's success. If the other urn isn't empty then, you've failed to generate a word in $T^{-1}t$. Then abort that attempt, and try again.

Example: For $t = (0; 1, 2, 1, 0)$ the two urns, and the graph, are composed as in Figure 7. (For the present, ignore the suggestion that one ball is stuck to the bottom of urn 0.) Since $i = 0$, w begins with a 0 and we draw its second letter from urn 0.

Fig. 7.

Case 1. Thc 0 is drawn, yielding 00 so far. Then the next draw is from urn 0, too, where only 1's remain: that's 001 so far. The next draw takes the 0 from urn 1 (0010, so far) and the last takes the remaining 1 from urn 0, leaving both urns empty. Success. Result: $w =$ 00101.

Case 2. A 1 is drawn: 01. Then the next draw empties urn 1 (010), and the next takes the 0 or the 1 from urn 0. There are two possibilities.

Case 2a. The 0 is taken: 0100. Then the last draw empties urn 0. Success. Result: $w = 01001$.

Case 2b. The 1 is taken: 0101. Then the next draw is to be from urn 1, which is empty – while a 0 remains in urn 1.

There are only two successful cases: 1 and 2a. For long enough runs of this process, your prior judgmental probability that those two cases will occur in a ratio between (say) 999:1000 and 1000:999 will be as close as you like to 1.

That was a crude version of the urn model. In fact, Diaconis and Freedman (1980b, p. 240) arrange matters so that you can't fail: you'll never be asked to draw from an empty urn. (The following vivid account of that modification was suggested by Diaconis. Zaman (1984), Section 7 extends the modification to the general case, of any finite number of letters.)

The Glued Ball Method (where j is the letter that isn't i). If $\#ij = \#ji$ in t, glue one of the i balls to the bottom of urn j; if $\#ij = 1 + \#ji$, glue one of the j balls to the bottom of urn i. Glued balls don't come out. If you are ever instructed to draw from an urn containing only a glued ball, use that ball's label as the word's last letter.

In the example, i was 0 and $\#ij$ was 2 while $\#ji$ was 1, so we glued a j ball to the bottom of urn i. In cases where no gluing is called for, no sequence of draws can fail; but where gluing is called for there are ways to fail, without it, e.g., case 2b above, which cannot arise when a ball is glued as in Figure 7. (With 010 as first three letters, we must draw from urn 0, where, since the "1" ball is glued, we must draw the 0, which puts us in the successful case, 2a.)

The process can equally well the described in graphical terms: see Figure 8, where the letters of w are to be the labels of the nodes you visit, starting at node i and always following arrows.

The Shaded Node Method (where $j \neq i$). Shade node i if $\#ij = \#ji$; shade node j if $\#ij = 1 + \#ji$. When you traverse an arrow, reduce its numerical label by 1; and erase arrows labelled 0. Never traverse an arrow labelled 1 from an unshaded node to a shaded one when you have another option. Success is ending with no arrows; failure is finding that there's one you can't traverse from where you've ended.

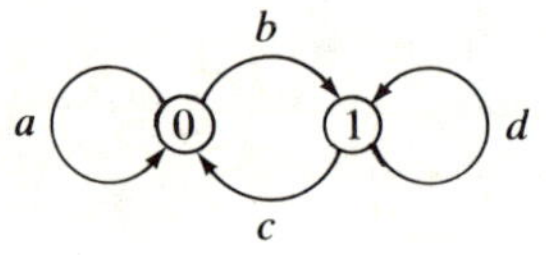

Fig. 8.

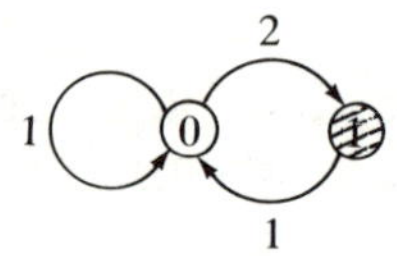

Fig. 9.

Example: $t = (0; 1, 2, 1, 0)$ again. Then the relevant graph is Figure 9, where node 1 is shaded for the same reason that a "1" ball was stuck to the bottom of urn 0 in Figure 7. Note that in cases 1 and 2a above, all arrows would eventually vanish even without the shading; and note that case 2b cannot now arise, for you may not enter the shaded node while you have the option of generating another 0 instead.

Let M_t be the probability distribution over L^* that corresponds to the urn or graph model described above, i.e., uniform on $T^{-1}t$ and 0 elsewhere in L^*. *Markov symmetry* or "partial" *exchangeability relative to a Markov tally function T* is a matter of uniformity (22) relative to T. Corresponding to (26) and (27) in the case of simple exchangeability we have

(32) $P(\ |T^{-1}t) = M_t$ for any Markov-symmetrical P.

(33) *De Finetti Representation for Markov-Symmetrical P.*

$P(\ |L^n) = \Sigma_t a_t M_t$ where $a_t = P(T = t|L^n)$ and t ranges over the tallies of words of length n.

Corresponding to (29) we have (Diaconis and Freedman 1980b):

(34) *De Finetti Representation, Infinite Case.* If P is Markov-Symmetrical on L^∞ and $P(0$ infinitely often$) = 1$, there is a distribution M on $X = [0, 1]^2$ relative to which P is a mixture of 2-state Markov chains:

$$P(\{w\} + L^\infty | \{i\} + L^\infty) = \int_X p^a(1 - p)^b q^c(1 - q)^d \, \mathrm{d}M$$

for all w in L^*, where $(i; a, b, c, d) = Tw$, and where P and q are the transition probabilities of Figure 10.

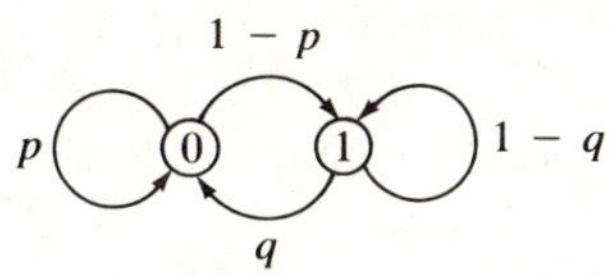

Fig. 10.

And corresponding to (30) we have the integral in (34) as an approximation to $P(\{w\} + L^* | \{i\} + L^{k-1})$ when k is much greater than the length of w.

PARTIAL EXCHANGEABILITY

Partial exchangeability is a genus, i.e., exchangeability relative to one or another tally function; simple and Markov exchangeability are two of the oldest and best-studied species in that genus. Others are described in Diaconis and Freedman's (1980b) excellent introduction to the whole topic, on which the foregoing account of Markov exchangeability is based, and in a later paper (1982), where they describe the genus in dazzling generality, survey the known species, and report on the state of the art.

The art is a matter of (1) finding de Finetti representations for different species of partial exchangeability, or of (2) going the other way around. In the case of simple exchangeability, task (1) starts with the characterization

(35) $\quad P\{x\} = P\{y\}$ if x is an anagram of y

and ends when the hypergeometric and multinomial distributions are identified as the extreme points in L^* and L^∞, various mixtures of which yield the various symmetrical distributions. But the thing may really have started in the opposite direction, (2), with de Finetti wondering how to characterize mixtures P of multinomial distributions in L^∞ directly, discovering that they are exactly the distributions satisfying (35), and only then wondering (1) about the extreme points in L^*, the different mixtures of which are the different P's satisfying (35). And as Zaman [1984] suggests, (2) is the much more likely direction in the case of Markov exchangeability.

Anyway, de Finetti representations give special interest to concepts of partial exchangeability, and, so, to tally functions. As (35) suggests, concepts of partial exchangeability can also be characterized by the groups of transformations under which the distributions in question are invariant – groups that define congruence relations on the sample space. (Here von Plato's contribution to the present volume is relevant.)

Even when a species of partial exchangeability is characterized by a tally function T, a further generalization is easy: the uniformity condition (22), "$P\{x\} = P\{y\}$ if $Tx = Ty$", can be dropped in favor of the weaker sufficiency condition (7), which it implies – in the following form (where $T^{-1}t$ is the proposition that T assumes the value t).

(36) For each t, conditioning on $T^{-1}t$ (the hypothesis that $T = t$) makes all distributions in the species identical.

Example: mixtures of Poisson distributions (Diaconis and Freedman 1980, p. 246–7). But the congruence condition is indispensable for the representation theorems. That is condition (31) here, or condition (1.2) in Diaconis and Freedman (1982), p. 207.

Sufficiency = partial exchangeability = applicability of probability kinematics. Then where we have a representation theorem for a species of partial exchangeability, probability kinematics can have a special sort of rationale.[18]

Example: the die may be loaded. Perhaps you think that someone who had all relevant information about the die, and had enough time and knowledge of physics to use it, would have a multinomial judgmental probability distribution on $L^* = \{1, 2, 3, 4, 5, 6\}^*$ in which the six possible outcomes of each toss need not be equiprobable. Under such circumstances you may well seek your own judgmental probability distribution among the mixtures of multinomial distributions.

By de Finetti's representation theorem, the statistic (21) $T = (N_1, \ldots, N_6)$ is then sufficient for the set of all candidates for the position of your judgmental probability distribution P over L^*. Note that a definite P may be more than you need, e.g., you may be content with the conditional distributions $P(\ |L^n)$.

CONCLUSION

Partial exchangeability and probability kinematics are a single thing seen from different practical perspectives.

In each case there is a statistic (tally function) T for which the conditional probabilities

$$P(\ |T^{-1}t)$$

play essentially the same, pivotal role. In kinematics these conditional probabilities remain valid for the new probability distribution. Thus the difference between the old and new distributions is determined by the probabilities they assign to T's sets of constancy. In the case of partial exchangeability these conditional probabilities are the extreme points, and every probability distribution that's partially exchangeable relative to T is a convex combination of extreme points.

There is no magic machine to produce useful statistics T, but there is a collection of examples that have proved useful here and there, e.g., familiar classical probability models and their sufficient statistics. All are grist for the mill, i.e., kinematics or T-exchangeability, depending on the point of view.

Consider the kinematical interpretation of the example discussed under Markov exchangeability above of a thumbtack that is flicked repeatedly, always from the position in which it landed. Many statisticians have treated this as a case of simple exchangeability. Reading the present analysis might give them pause: things *aren't* exchangeable if the last toss matters. If that's all that matters then the tally T of the number of transitions (0 to 0, 0 to 1, 1 to 0, 1 to 1) is a sufficient statistic. Here the uniform conditional probability $P(\ |T^{-1}t)$ specifies P except for four numbers, which must be thought about and specified directly. The kinematical change comes through introspection and insight.

There may be other partial exchangeabilities that are worth recycling as kinematics (e.g., among those surveyed in Diaconis and Freedman, 1982).

Probability kinematics and partial exchangeability have different ends in view. Kinematics aims at sensible changes in belief, making full use of relevant past experience. Partial exchangeability is a subjectivistic version of model building.

Both can draw on the same rich lore of applied statistics, accumulated over the years, for ideas about useful statistics T.

Department of Philosophy
Princeton University

NOTES

Thanks are due to the National Science Foundation and Princeton University for support of this work, and to Persi Diaconis, David Freedman, David Lewis, Brian Skyrms, Bas van Fraassen, and Sandy Zabell for lots of help. (The conclusion is actually by Diaconis, somewhat rewritten by Jeffrey.)

[1] Also known as "conditionalization".

[2] Lewis's argument is reported in Teller (1976), section 1.3. Freedman and Purves (1969) have a related but different result.

[3] The present restriction to countable partitions is inessential: see Jeffrey (1965, 1983) section 11.8 for informal discussion, and Diaconis and Zabell (1982), section 6, for a rigorous treatment of the general case.

[4] Diaconis and Zabell (1982), p. 824, (2.2), point out that the connection of (9) with sufficiency is a version of the Fisher–Neyman factorization theorem.

[5] This proof is from van Fraassen (1980), p. 173. Diaconis and Zabell (1982), p. 824, point out a proof in Blackwell and Girshick (1954), p. 221.

[6] If D is to be a subset of the sample space (one to which P assigns a value), Q must be a random variable: each sample point w must specify a complete hypothesis Q_w about what your new probability function will be. Then $D = \{w : Q_w(E_i) = q_i, i = 1, \ldots, n\}$. An unknown sample point t represents reality (truth), and your actual new probability function will be Q_t. But think of the "Q"s in the text, e.g., in formula (8), as denoting Q_t, your actual new probability distribution, just as "The population of Chicago in 1900" denotes a number, not a function; not a random variable.

[7] Skyrms (1980a, appendix 2, and 1980b, pp. 123–7). I. J. Good (1981) makes a similar suggestion (which he told me about in 1963). The earliest such proposal that I know of, by Jaynes (1959) pp. 152–164, "The A_p Distribution: Memory Storage for Old Robots", was independent of probability kinematics.

[8] Skyrms (1980a), p. 125.

[9] Corresponding to our e_i Field uses $\alpha_i = \log e_i + M$, where M is the constant $(1/m)\Sigma_i \log[q_i/P(E_i)]$.

[10] See sections 3 and 4 of Diaconis and Zabell (1982) for details.
[11] In correspondence reproduced in Jeffrey (1975).
[12] We could mark the present by using ordered pairs of words as scenarios, with the first member representing what has already come to pass and the second representing what still lies ahead. In this notation $\langle w, o\rangle$ is the form of scenarios in which the sequence of tosses has ended, and $\{\langle w + 1, x\rangle : w, x \text{ in } W\}$ is the proposition that there have already been one or more tosses, the most recent of which yielded an ace.
[13] This treatment of exchangeability is based on Diaconis (1977). See also Skyrms (1984), chapter 3.
[14] Here "P" denotes distinct probability distributions over two different spaces, L^∞ and L^*, in which the proposition that the sequence of outcomes begins with w is represented by the sets of $\{w\} + L^\infty$ and $\{w\} + L^*$, respectively.
[15] Diaconis (1977) and Diaconis and Freedman (1980a) prove this theorem and derive the infinite form as a corollary, and point out other proofs of finite forms of the theorem by de Finetti (1972, p. 213) and others.
[16] See Freedman (1977): "When drawing a sample of 1000 from a population of 100 000 000, there is almost no difference between drawing with or without replacement [i.e., the error is between 0.00498 and 0.00500]. When drawing a sample of 5000, there is a substantial difference in variation norm [i.e., the error is between 0.117 and 0.125]."
[17] Condition (31) here is a translation into the L^* idiom of Diaconis and Freedman's (1982, p. 207) condition (1.2). It is needed in order to prove the representation theorem in the infinite case.
[18] For various examples see Diaconis and Zabell (1986).

REFERENCES

Achinstein, Peter and Owen Hannaway (eds.) (1985) *Observation, Experiment, and Hypothesis in Modern Physical Science* (Cambridge, Mass.: Bradford-MIT Press).

Bernardo, J. M., M. H. DeGroot, D. V. Lindley, and A. F. M. Smith (eds.) (1985) *Bayesian Statistics 2* (Amsterdam: Elsevier, North-Holland).

Blackwell, David and M. A. Girshick (1954) *Theory of Games and Statistical Decisions* (New York: Wiley).

Carnap, Rudolf (1950) *Logical Foundations of Probability* (University of Chicago Press; 2nd ed., 1962).

de Finetti, Bruno (1937) 'La prévision: ses lois logiques, ses sources subjectives', *Annales de l'Institut Henri Poincaré* **7**, 1–38. Translated in Kyburg and Smokler (1964, 1980).

de Finetti, Bruno (1972) *Probability, Induction and Statistics* (New York: Wiley).

de Finetti, Bruno (1938) 'Sur la condition d'équivalence partielle', *Actualités Scientifiques et Industrielles*, No. 739 (Paris: Hermann & Cie.) Translated in Jeffrey (1980).

Diaconis, Persi (1977) 'Finite forms of de Finetti's theorem on exchangeability', *Synthese* **36**, 271–281.

Diaconis, Persi and David Freedman (1980a) 'De Finetti's generalizations of exchangeability', Jeffrey (1980), 233–249.

Diaconis, Persi and David Freedman (1980b) 'Finite exchangeable sequences', *Annals of Prob.* **8**, 745–764.

Diaconis, Persi and David Freedman (1982) 'Partial exchangeability and sufficiency', Proceedings of the Indian Statistical Institute Golden Jubilee International Conference on *Statistics: Applications and New Directions*, Calcutta, 16–19 December, 1981, pp. 205–236 (Calcutta: Indian Statistical Institute).

Diaconis, Persi and Sandy Zabell (1982) 'Updating subjective probability', *J. Am. Stat. Assn.* **77**, 822–830.

Diaconis, Persi and Sandy Zabell (1986) 'Some alternatives to Bayes's rule', in Grofman and Owen, pp. 25–38.

Field, Hartry (1978) 'A note on Jeffrey conditionalization', *Philosophy of Science* **45**, 361–367.

Freedman, David (1977) 'A remark on the difference between sampling with the without replacement', *J. Am. Stat. Assn.* **72**, 681.

Freedman, David and Roger A. Purves (1969) 'Bayes' method for bookies', *Annals of Math. Stat.* **40**, 1177–1186.

Garber, Daniel (1980) 'Field and Jeffrey conditionalization', *Philosophy of Science* **47**, 142–145.

Goldstein, Michael (1983) 'The prevision of a prevision', *J. Am. Stat. Assn.* **78**, 817–819.

Goldstein, Michael (1985) 'Temporal coherence', in Bernardo *et al.*, 231–248.

Good, I. J. (1981) 'The weight of evidence provided by uncertain testimony or from an uncertain event', *Journal of Statistical Computation and Simulation* **13**, 56–60.

Grofman, Bernard and Guillermo Owen (eds.) (1986) *Information Processing and Group Decision Making* (Greenwich, Conn.: JAI Press).

Harper, William and C. A. Hooker (eds.) (1976) *Foundations of Probability Theory, Statistical Inference, and Statistical Theories of Science* (Dordrecht: D. Reidel Pub. Co.).

Jaynes, E. T. (1959) *Probability Theory in Science and Engineering* (mimeographed) (Dallas: Field Research Laboratory, Socony Mobil Oil Co.).

Jeffrey, Richard (1957) *Contributions to the Theory of Inductive Probability* (Ph. D. Dissertation, Princeton University).

Jeffrey, Richard (1965) *The Logic of Decision* (New York: McGraw-Hill Pub. Co.: 2nd edition, University of Chicago Press, 1983).

Jeffrey, Richard (1975) 'Carnap's empiricism', in Maxwell and Anderson (1975).

Jeffrey, Richard (ed.) (1980) *Studies in Inductive Logic and Probability*, vol 2 (Berkeley and Los Angeles: University of California Press).

Jeffrey, Richard (1985) 'Probability and the Art of Judgment', in Achinstein and Hannaway (1985), 95–126.

Kyburg, Henry E., Jr. and Howard E. Smokler (eds.) (1964) *Studies in Subjective Probability* (New York: Wiley. 2nd ed., Robert E. Krieger: Huntington, New York, 1980).

Lewis, C. I. (1946) *An Analysis of Knowledge and Valuation* (La Salle, Ill.: Open Court).

Maxwell, Grover and Robert M. Anderson (eds.) (1975) *Induction, Probability, and Confirmation* (Minneapolis: University of Minnesota Press).

Mellor, D. H. (ed.) (1980) *Prospects for Pragmatism* (Cambridge: The University Press).

Skyrms, Brian (1980a) *Causal Necessity* (New Haven: Yale University Press).

Skyrms, Brian (1980b) 'Higher-order degrees of belief', in Mellor (1980), 107–137.

Skyrms, Brian (1984) *Pragmatics and Empiricism* (New Haven: Yale University Press).

Skyrms, Brian (1987) 'The value of knowledge', in *Justification, Discovery and the Evolution of Scientific Theories*, C. Wade Savage (ed.), (Minneopolis: University of Minnesota Press).

Suppes, Patrick (ed.) (1980) *Studies in the Foundations of Quantum Mechanics* (East Lansing, Michigan: Philosophy of Science Association).

Suppes, Patrick and Mario Zanotti (1980) 'A new proof of the impossibility of hidden variables using the principles of exchangeability and identity of conditional distributions', in Suppes (ed.), 173–191.

Teller, Paul (1976) 'Conditionalization, observation, and change of preference', in Harper and Hooker (1976), pp. 205–259.

van Fraassen, Bas (1980) 'Rational belief and probability kinematics', *Philosophy of Science* **47**, 165–187.

van Fraassen, Bas (1984) 'Belief and the will', *J. Phil.* **81**, 235–256.

von Mises, Richard (1919) 'Grundlagen der Wahrscheinlichkeitsrechnung', *Mathematische Zeitschrift* **5**, 52–99.

von Mises (1957) *Probability, Statistics, and Truth*, 2nd rev. ed. (London: George Allen and Unwin. Dover reprint, 1981).

Zaman, Arif (1984) 'Urn models for Markov exchangeability', *Annals of Probability* **11**, 223–229.

JAN VON PLATO

ERGODIC THEORY AND THE FOUNDATIONS OF PROBABILITY

1. INTRODUCTION

In the following, I shall illustrate through examples how useful the notions and results of ergodic theory are for the foundations of probability. Ergodic theory can be discussed from many points of view. It originated with the development of statistical mechanics in classical physics. Later, an abstract formulation of ergodic theory became part of measure theory. The latter treats probabilities as special kinds of measures. Results of ergodic theory can be intuitively explained in probabilistic terms, without special recourse to their physical significance or to the uninterpreted, abstract mathematical formulation. In discussing the relevance of ergodic theory for the foundations of probability, the physical background is significant, however. Most of the notions of the theory are motivated from statistical physics, so that, proceeding in the other direction as it were, an interpretation of the results of ergodic theory can be given in physical terms. It should be noted that there are different possibilities to hand here. Some writers consider the physical application a special case of *a priori* probabilities.[1] Against this view, in which probabilities only relate to ignorance, stand the attempts at justifying probabilistic assumptions from a physical basis. Below, we will specifically show that there is a reason why *uniform a priori* probability distributions yield physically correct results. On the other hand, if the criterion proposed is not fulfilled, they cannot give such results.

In what follows, the basic concepts are first explained in terms of a state space of a system which is adequate for classical statistical mechanics. This will be a subset of a real space R^n. Afterwards, a general formulation is given.

There exists an interpretation of probability that has been common in physics during all of this century, but almost impossible to find in researches on foundations of probability. It is called the *time average* interpretation of probability. Its inventor is Ludwig Boltzmann, in 1868, and it was also discovered by Einstein in his earliest works.[2]

Brian Skyrms and William L. Harper (eds.), Causation, Chance, and Credence, Vol. I, 257—277.

Consider an arrangement or event sometimes taking place, sometimes not. The relative time that the arrangement is taking place is well defined for any given finite period. If there is a limit to this relative time as the period goes to infinity, and if that limit does not depend on the particular circumstances at hand but is in general unique, it equals the probability of the event interpreted as limit of time average. As indicated, the limits do not come about by themselves, but need special conditions. These conditions will also tell us when the limits are unique, or independent of the specific actual course of events. With discrete time, limits of time averages become limits of relative frequencies. 'Time averagist' probability is not the same as frequentist probability, however, but a more specific notion.

A rival of the time average conception is what is in physics called the ensemble interpretation of probability. In what is known as the microcanonical case, a distribution is postulated over the state variables of a system. Now, instead of only one system, a continuous collection of systems is imagined such that each specific system appears in the collection with a density equalling the probability. It is not always clear what the intended interpretation of the ensemble is.[3] One view, mentioned already above, is that it is just an imaginary device, reflecting how well we have been able to identify the actual system in front of us. This is the aprioristic interpretation of ensemble densities. Another is one in which a random choice from the ensemble is postulated. In a discrete formulation, the difference with the time average approach can be likened to tossing an infinite number of coins at one time against tossing one coin an infinite number of times.

The illustrations to be presented are as follows.[4] First, it is shown that the notion of ergodicity gives us a *characterization* of cases where probability as time average, and frequentist probability more generally, can be applied. Familiar laws of large numbers only give a sufficient condition (independence and identical distribution) and therefore leave out relevant cases. Secondly, it is shown that ergodicity and stationarity are related in exactly the same way as the notions of independence and exchangeability. The two latter stand in a relation shown by de Finetti's representation theorem. This fits very nicely to the first point. Putting ergodicity, the characteristic of unique limits of relative frequencies, in place of independence, will give us a "best possible" case of a generalization of de Finetti's theorem. Our third point gives the condition for introducing uniform probability distributions, and is again a form of ergodicity criterion expressed as a condition for the dynamics of a system. The fourth point studies the case where this condition is not fulfilled. A theoretical scheme is suggested for discussing scientific experimentation in which such notions as causality, causal factors, probabilistic causality and explanation can be given precise meaning.

2. CHARACTERIZATION OF FREQUENTIST PROBABILITY

I will first introduce the subject up to a point where the central result of the field, the *ergodic theorem* of Birkhoff, can be presented. Its abstract formulation is the proper optimal generalization of the strong law of large numbers. A comparison of time average and relative frequency is indicated, and finally it will be seen in what sense probabilities can be *identified* with limits of time averages.

The following notions are borrowed directly from the classical dynamics of point particles. There will be a *state space*, also called phase space, X. Typically $X \subset R^n$, for example, a simplest case would have one coordinate for position and one for velocity (or momentum), so that the state space would be a subset of the Cartesian plane R^2. There will be constraints on the system such that the set of physically accessible states is a connected set of finite extension (area, volume etc.). The motion of a *microstate* $x \in X$ is governed by the *dynamical law* of the system. The following assumption will be made: It is required that the dynamical law is independent of time. This means that if the system is in state x at time t_1, or at time t_2, the state x will evolve in exactly the same manner. Time is represented by a real number. For each x and t, $T_t x$ will denote the state to which the system evolves in time t, starting from state x. The continuous collection of transformations $\{T_t\}_{t \in R}$ will form a *dynamical group* with a real parameter. The operation with respect to which it has the group property is that of combining transformations: there will be, for each t_1 and t_2, a unique transformation which takes x to $T_{t_1}x$ to $T_{t_2}(T_{t_1}x)$, and it is $T_{t_1+t_2}x$. Inverses are defined by $T_t^{-1} \equiv T_{-t}$.

The microstate x determines whatever physical properties the system has at the time it is in state x. We suppose that all physical properties are identifiable as functions defined over microstates, called phase functions. In the other direction, it is conceivable that not all phase functions represent physically meaningful properties. A classic proposal is the function which takes value 0 on coordinates having rational values, and 1 elsewhere.[5] The time average of a phase function f from t_1 to t_2 is its average value when the system evolves from state x at time t_1 to state $T_{t_2-t_1}x$ in the time interval $t_2 - t_1$. As the average is over a continuous number of values of f, it is an integral $\int_{t_1}^{t_2} f(T_t x)\,dt$. Relative time is the value of the integral divided by the length of the interval $t_2 - t_1$. Now, it can be shown[6] that for a system with time independent dynamics, the limit

$$\lim_{t\to\infty} t^{-1}\int_0^t f(T_t x)\,dt = \hat{f}(x)$$

exists. The limit would be the same if another value were substituted for zero. Notice that the value of the limit is a function of x here. Next, let us call the set $\{y \mid T_t x = y, t \in R\}$ the *trajectory* or orbit of x. A set A is a *complete set of trajectories* if for each $x \in A$, it contains the trajectory of x. There are no trajectories leading "in" or "out" of A in this case. If all complete sets of trajectories are of essentially two kinds, either essentially the same as the whole state space X or else essentially empty, the system is *ergodic*. By "essentially" is meant the following. X is a connected subset of finite extension of a real space R^n. The generalization of geometric measure, Lebesgue measure m gives the same measure to two sets whose difference has measure zero. In these terms, ergodicity requires complete sets to be of the same measure as the whole state space X, or of measure zero. Further, the condition of time independence of a dynamical law suggests that each of the T_t should be a *measure preserving transformation:* $m(T_t^{-1}A) = m(A)$. Here, as is usual, one requires that the T_t are measurable functions with respect to m and $T_tA = \{T_t x \mid x \in A\}$, $T_t^{-1}A = \{x \mid T_t x \in A\}$. If A is the range of some set under the mapping T_t, the inverse image of A, $T_t^{-1}A$ has to be measurable with respect to m, i.e., $m(T_t^{-1}A)$ has to be defined. In this way, the measure m yields values for sets generated by T_t from arbitrary measurable sets. In the present case, the inverse T_t^{-1} is not a one-to-many relation but a function. The property of measure-preservedness can be written as $m(T_tA) = m(A)$. Anticipating things below, let us call systems preserving measure *stationary*. Next, call *invariant* sets which remain the same under the transformations T_t: $T_t^{-1}A = A = T_tA$. A second formulation of the condition of ergodicity is that all invariant sets must have either measure zero or one.

The central result of the field, the ergodic theorem of Birkhoff, says that limits of time averages of phase functions are essentially unique exactly in the case that the system is ergodic. Then, $\hat{f}(x)$ does not depend on x but is a constant $\hat{f}$. Its constancy does not hold over all states (or trajectories), but the exceptions have measure zero in terms of the Lebesgue measure referred to above. Further, if the Lebesgue measure m over X is normalized (if not, it will suffice to divide $m(A)$ by $m(X)$), $\hat{f}$ can be determined as an average over the whole of X. It is the *phase average* $\bar{f}$ of f over $X: \bar{f} = \int_X f(x)\,\mathrm{d}m$. The ergodic theorem can be stated as $\bar{f} = \hat{f}$, or time average equals phase average. The original problem of ergodic theory in statistical mechanics was a determination of time averages, and it is thus achieved if the condition of ergodicity is fulfilled. It was thought that finite time averages might approximate the limits, so that one would be able to predict the macroscopic behaviour of statistical mechanical systems at least for sufficiently long time intervals. In a somewhat different interpretation, it is accepted that finite time averages may fluctuate. Instead, the prediction concerns the long run average of these varying finite time averages.[7]

Probability enters the picture in the following way. Let I_A be the indicator function (characteristic function) of the set $A \subset X$. Under ergodicity, the limit of time average of I_A is unique and equals its phase average. The latter is the same as the expected value of I_A relative to the normalized Lebesgue measure m, or in other words, the probability of A.

Let us next look at a more general formulation, the structure of which is rather directly obtained from the above. There will be an abstract space X with a σ-algebra of events $\mathscr{A}$ and, in the simplest case, one transformation T together with its iterates $T^2, T^3, \ldots$. There will be a normalized measure m over X. It is required that T is a *measure preserving* transformation with respect to m: the inverse image $T^{-1}A$ of any measurable subset $A \in \mathscr{A}$ of X should be of the same measure as $A: m(T^{-1}A) = m(A)$. The motivation for this requirement was indicated above: If the dynamical law of a classical system is independent of time, the motion of the system will give a transformation which is measure preserving with respect to the natural measure m above. That is, $m(T_t^{-1}A) = m(A)$. This is known as Liouville's theorem and can be found in almost any presentation of equilibrium statistical mechanics. In the abstract formulation, it appears as a postulate. The consequence of this postulate is similar to one presented already. In the formulation with one transformation and its iterates, we have for the indicator function that

$$\lim_{n\to\infty} n^{-1} \sum_{i=1}^{n} I_A(T^i x)$$

exists. That is, the limit of relative frequency of occurrences of the event A in the sequence x, Tx, T^2x, ... exists. There is a set of exceptional sequences but it has m-measure zero. A set A is invariant if its inverse image under T equals $A: T^{-1}A = A$. If all invariant sets have either m-measure zero or one ($m(A) = 0$ or 1), T is called ergodic relative to m. It follows that limits of relative frequencies have unique values, except for a set of m-measure zero. These values coincide with the probabilities. Furthermore, the situation is optimal in the sense that the uniqueness of limits of relative frequencies is characteristic for ergodicity. For this reason, ergodicity should be the central notion in at least those frequentist interpretations of probability which accept the common measure theoretic framework.

It would be possible to give the abstract formulation in terms of a group of transformations such as $\{T_t\}$ in the classical case. The ergodic theorem would read exactly as it stands above for classical systems.

Let us now turn into a discussion of the significance of these results. The existence of limits of time averages was in the classical case guaranteed by the invariance of the laws of motion in time. For statistical mechanical systems, this is usually expressed as the requirement that the total energy of the system is constant in time, so that the system does not interact physically with its surroundings. Another formulation was the condition of stationarity. Ergodicity requires that in addition to this, the set of trajectories of the system must not form invariant subsets having measure strictly between zero and one. Therefore, both conditions depend on properties of the dynamical law of the system. For classical systems, there is a unique normalized measure m which is preserved under the dynamical motion. It is defined, if the dynamics is given in terms of Hamiltonian canonical coordinates q_1, ..., p_n, by the differential volume element $dq_1 \ldots dp_n$. The same measure appears in the condition of ergodicity. So far everything has been formulated in terms of dynamical quantities. It has, indeed, been claimed that the whole programme of ergodic theory in classical dynamical systems can be formulated with no recourse to probabilistic notions.[8] However, the set of trajectories for which time averages exist (or the set for which they are unique) stands, without further qualification, on a par with lots of other continuous sets of trajectories having some property or other. The empirical meaning of the measure giving such a special status to some sets rather than others, may be gathered from the following consideration. Suppose we were given the task of preparing, say repeatedly, the same ergodic system, and of observing its statistical behaviour (the time averages). To violate against what the theory predicts, we should be able to prepare the system into initial states having m-measure zero. Each sequence of preparations has of course measure zero because it is at most countable. However, "ability" is not modelled by one accidental sequence of measure zero, but by a procedure. If preparation with absolute exactness is excluded as physically impossible, the preparation procedure involves a statistical element. This is represented by a probability distribution for preparing the state. To violate the statistical predictions, this probability should give positive measure to sets of states having m-measure zero.[9] In fact, as will be shown in section five, the situation can be seen as a violation of the original ergodicity assumption of the system. The possibility of a preparation of the indicated kind will be seen to involve the existence of a causal factor affecting the right physical description of the situation.

In going over to the abstract formulation, several things happen. The dynamical law no more leads to the conditions of stationarity or ergodicity, but these appear as postulates or probabilistic assumptions. One usually calls the measures or probabilities stationary and ergodic here, instead of the transformation T. If the probabilistic models represent repetitive experimentation, the role of T is to 'give the next result'.

Frequentist probability used to be based, in the theory of von Mises for example, on two assumptions: first, that there is a random choice of the repetitive events, and, secondly that relative frequencies have unique limits. The criterion of random choice can be defined as insensitivity of the limit with respect to systematic choices of subsequences from the events. A typical way to model these in present probability theory is to postulate probabilistic independence (and identical distribution) between consecutive events or "trials", so as to meet condition one. The second follows by the law of large numbers. Measure theoretic formulations of von Mises' criteria of randomness were given in the thirties by Doob and Halmos. The requirement implies that probability on denumerable sequences of events $x = (x_1, x_2, \ldots)$ should be invariant with respect to transformations $T_f x = (x_{f(1)}, x_{f(2)}, \ldots)$ where f is a strictly increasing function $f: N \rightarrow N$. Halmos shows that such transformations are measure preserving for independent, identically distributed probability measures, and says that this result corresponds to von Mises' idea of randomness.[10] However, it should be noted that it is a special case: it is known (*cf.* Kingman 1978) that the condition of exchangeability can be formulated equivalently as follows. It is required that $P(x_1, x_2, \ldots) = P(x_{n_1}, x_{n_2}, \ldots)$ where $1 < n_1 < n_2 < \ldots$, or in other words, invariance with respect to strictly increasing subsequences is an equivalent condition. Independence is, in fact, a special case of ergodicity of a probability measure, and the law of large numbers a special case of the ergodic theorem. In the classical case, the two desirable properties of frequentist probability appear as consequences of the condition of ergodicity and related conditions on the dynamical law. Here, a criterion is sought for fulfilling the second condition on frequentist probability. Under ergodicity, it is not possible to select a subset of positive measure of the state space that would lead to different limits of time averages. One could say that sampling or repetition is performed by the dynamics, and ergodicity guarantees that it is also fair because one always gets the same statistical behaviour. This relates to von Mises' formulation of the requirement of randomness, namely that there should be no way of choosing a subsequence from the sequence of repetitions which would bring us different limiting frequencies.[11] As ergodicity is an asymptotic property, selection could maintain correlations for any finite times. Stronger conditions are needed if this is to be avoided.[12] All these can be called *conditions of randomness.*[13]

Our conclusion is that probability as time average solves, in a special case, two outstanding problems of frequentist probability. First, the problem of what one means by a homogeneous population, also known as the problem of reference class. Almost all samples turn out fair for the proposed criterion of homogeneity. Secondly, it solves the question of the existence and uniqueness of frequentist probabilities. The price paid for these advances is, at least in the present state of science, that there are very few cases in which dynamics is well enough determined (or mastered) to yield ergodic theorems. These few cases may even appear hopelessly unrealistic, absolutely round and rigid balls in a rectangular container and so on, with a wrong (=classical) account of the microscopic scale of things. Interesting cases might nevertheless be offered by fields where probabilistic methods combine with good theoretical understanding of the basic interactions, such as population dynamics with molecular genetics as its basis.

It is naturally an idealization that there should be systems that do not interact with their surroundings. However, one often encounters cases where the ideal character of isolatedness seems inessential. One does not think that the equilibrium properties of statistical mechanical systems depend critically on absolute isolatedness. More generally, if we want to think in terms of stable statistical properties in the sense of unique limiting frequencies, admitting at the same time the ideal character of this assumption, we are unavoidably led to the probabilistic form of the notion of ergodicity. If the stable statistical properties are explained by probabilistic laws of nature, a justification of these laws analogous to the case of classical ergodicity should be possible from first principles.

3. STATIONARITY AND ERGODICITY RELATE LIKE EXCHANGEABILITY AND INDEPENDENCE

Probabilistic independence has always been one of the central notions on which frequentist or objectivist interpretations rest. If an experiment is performed repetitively in the same circumstances, it is supposed that each time the same probability distribution holds for the different possible results. If consecutive events do not affect each other, independence is seen as fulfilled, and the law of large numbers can be applied. An intuitive notion of causal independence is used which requires that no memory or trace of the previous event is left over when the performance of the experiment is prepared for.

Our view is that identical circumstances in the repetition of an experiment only suffice for introducing stationary probabilities.[14] This is a strong, positive statement even if it here appears as a denial of independence. Any experimental arrangement, if not altered, produces stationary sequences of events. Relative frequencies converge.

It has indeed been usual to infer probabilistic independence from causal independence. The above is a denial of the said inference: If causal independence requires that an event in no way alters the circumstances in which another event occurs, it can still happen that the events are probabilistically correlated. It is, in fact, easy to think of correlated events with space-like separation.

Even if there should be no general recipes for it, let us for a while accept that in one way or other, one can get convinced about having identically distributed independent random variables as descriptive of certain repetitive experiments. It is precisely the situation which is challenged by subjectivist probabilists of de Finettian persuasion.[15] de Finetti claims, on the basis of his *representation theorem*, that there is nothing an objectivist can do, with the notion of independence, that he could not do, or do in a philosophically more satisfactory way, with his own notion of *exchangeability*. There are several formulations of it. The basic idea is that the *order* of results in experiments is irrelevant in the calculation of probabilities. A sequence of random variables $x_1, \ldots, x_n$ is exchangeable if the probability $P(x_1, \ldots, x_n)$ is invariant under permutations of its arguments. An infinite sequence $x = x_1, x_2, \ldots$ is exchangeable if its finite segments $x_1, \ldots, x_n$ are exchangeable for all n. The invariances of an exchangeable sequence can be represented in a conventional algebraic way: as a *group of invariance transformations*. These are exactly the permutations for each n, of the sequence $x_1, \ldots, x_n$. They are measure preserving with respect to P. Now, let us take the simplest case of a simple experiment with two possible outcomes, $x_i = 1$ or 0 (success or failure, occurrence or non-occurrence of the event etc.). Exchangeability requires independence of the order of successes, so that the probability is already determined by the number of experiments n and successes k. There are $\binom{n}{k}$ such sequences, so that conditional on k, $P(x_1, \ldots, x_n | k) = 1/\binom{n}{k}$. We now come to a useful formulation of exchangeability. For each n, the number of successes $\Sigma_{i=1}^{n} x_i$ is a *sufficient statistic*, that is, the distribution of $P(x_1, \ldots, x_n | k)$, is independent of $x_1, \ldots, x_n$. k is an exhaustive summary of the data as far as the values of P are concerned. We are now able to state the central result about exchangeability, de Finetti's representation theorem: The probability $\omega_k^{(n)}$ of a sequence with k successes in n, if $x = x_1, x_2, \ldots$ is exchangeable for each n, is $\omega_k^{(n)} = \int_0^1 p^k (1-p)^{n-k} \, \mathrm{d}F(p)$ where F is a unique probability distribution over p. If F admits of a density, $\mathrm{d}F(p)$ can be written more familiarly as $f(p)\,\mathrm{d}p$ with f as the probability density over p. On the other hand, if one calculates probabilities as is done under the integral sign, and averages the probabilities with weights given by F, the result is an exchangeable probability. de Finetti's explanation of the result is as follows. The idea of constant objective (but unknown) probabilities p for independent repetitive experiments is illusory, but can be reconstructed on the basis of subjective probability. Moreover, there is nothing one could do with the objective probabilities that would not be accessible to the subjectivist. The left side $\omega_k^{(n)}$ of the equality is a subjective probability, a degree of belief in getting a sequence with k in n. The right side contains the illusory objective probability of success p, so that the illusory objective probability of the whole sequence is $p^k(1 - p)^{n-k}$. For given ω's, there is only one way of weighing the illusory objective probabilities, namely $F(p)$, which gives those ω's as an average. If one can do something with the scheme of unknown objective probabilities of independent trials, one could as well do it under the integral sign. de Finetti thinks the scheme is an unnecessary piece of metaphysics. One undesirable feature of it is that conditionalization on observations is trivial. The probability of a further success is unaffected by previous ones. This makes it impossible to "learn from experience", i.e., to adjust one's subjective probabilities according to one's experience of the world.

It is well known that exchangeable measures are stationary. Here stationarity appears in its standard probabilistic formulation: $P(x_{i_1}, \ldots, x_{i_n}) = P(x_{i_1+m}, \ldots, x_{i_n+m})$. That is, the unconditional probability of a sequence of length n is the same as the unconditional probability of any other sequence of length n. If one thinks of the index as giving the discrete time of the experiment with result x_i, stationarity in the above sense states that probability is invariant in time. It makes no difference when a series of experiments is begun. Now, the objectivist scheme of independence also is stationary, moreover, it is ergodic. Stationarity already gave us, for the case of exchangeability, the result that limits of relative frequencies exist but for a set of sequences of measure zero. For the case of independence, the law of large numbers assures us of the uniqueness of these limits[16] so that, as that uniqueness was stated to be characteristic of ergodicity, independence is a special case of ergodicity.

It is certainly natural to ask for a converse of what one gets with a law of large numbers. What is a necessary condition for the uniqueness of limits of relative frequencies? This was answered in Section 2 above, and the answer was just repeated. As the notion of independence is so central to de Finetti's philosophical reduction of objective probabilities to the subjective ones, one naturally asks for a necessary condition for the reduction, too. More specifically, what is the most general situation in which a probability interpreted as subjective is uniquely representable as a mixture of probabilities naturally interpretable as objective? Now, I took independence above as a way to unique limiting frequencies, and the latter situation certainly is one good objective for objective probability. Stationarity is equally good food for the subjectivist, especially if the crucial role of conditionalization is remembered: It characterizes the case where one's subjective probabilities remain the same in time as long as nothing is observed, and change only through conditionalization. A subjectivist philosopher of probability requires that probabilities are no esoteric properties of the external world.

They must be readily assessible. I think it is equally easy (or difficult) to decide if a sequence of events is symmetric with respect to order than if it is stationary. These two properties seem to stand on a par as subjective criteria. It is helpful to be able to think of them in terms of coordinate transformations: finite permutations or arbitrary increasing functions of indices for exchangeability, shifting indices an arbitrary number of steps for stationarity. Probabilities are invariant under arbitrary transformations for independence. I do not know how to formulate the condition of ergodicity in a similar manner. It would be helpful in choosing a class of probability distributions for a given experimental setting.

After the above preliminaries, it only remains to repeat what was already said in the title of this section: stationarity and ergodicity relate just like exchangeability and independence. If $P_S(A)$ is a stationary probability of an event A, there is a unique measure F over the collection of ergodic probabilities P_E such that $P_S(A) = \int P_E(A)\,\mathrm{d}F$. The rule is: Calculate the probabilities as if you had an ergodic measure, $P_E(A)$, then mix over the different possible cases with F.

The above result is known as the ergodic decomposition of stationary measures. It can be found in some advanced texts on probability. It was originally found by von Neumann, for the case of classical dynamical systems.

The present writer's philosophy is that one should not fix one's interpretation of probabilities once and for all. Rather, different interpretations are needed for the variety of cases arising in probabilistic everyday life. One step away from de Finetti's position is to allow for two kinds of probabilities, subjective and objective. de Finetti's theorem would say that as the objective probabilities are unknown, the possible values are weighed subjectively which makes the mixture (the exchangeable probabilities) subjective, too. With the ergodic decomposition, what is unknown are the limits of relative frequencies. One case where this kind of position has been taken is with classical systems. Stationary systems fall into ergodic components, and the trajectories are unable to pass from one to another. If it is unknown in which component we are (or the system is), time averages are unknown. A mixture over the components represents our uncertainty as to what the component is. Our pluralistic attitude is further vindicated by cases where no subjective weights seem to appear. Instead, the weights of the mixture are determined by physical quantities such as volumes of substances. The following would be an example. It is known that different phases of matter, such as liquid and gas, can coexist under certain circumstances. Furthermore, no physical mixing of phases occurs, but gas remains gas, liquid remains liquid and so on. Let us consider a can of coke. It is filled with coke-molecules, the chemical composition of which is a trade secret as everyone knows. Statistical physics tells us that there is a certain well defined temperature at which part of the coke is liquid, part in the form of coke-gas. It tells us also that there is a certain probability law for the particle density of molecules of the liquid, and another for those of the gas. The probabilistic law for the particle density distribution of all the coke-molecules in the can is a mixture of the above two, with weights given by the volumes of the two phases. The generalizations of de Finetti's representation theorem bring us cases where probabilities are not subjective in de Finetti's sense. In more general terms, we have a population which stratifies (to use Fisher's term) into several parts, each of which has its own statistical properties. Those of the whole population are represented as a mixture of the components' properties, with relative population sizes as weights. The properties of the whole are as objective as the properties of the parts.

4. WHY UNIFORM PROBABILITY DISTRIBUTIONS WORK?

Consider a classical system with state space X. Functions defined over X are called phase functions. If a phase function is a constant along the trajectory of x, so that $f(x) = f(T_t x)$, it is a *constant of motion* (invariant of motion, integral of motion). Let us suppose that our system is stationary. This can be expressed by the requirement that the total energy $H(x)$ of the system is constant over X, so that H is also a constant of motion. Constants of motion correspond to invariant sets and *vice versa*. If f is a (let us say real valued) constant of motion, define sets of $\{f^{-1}(x)\}$ by $\{y|f(x) = f(y)\}$. These sets are obviously invariant as they contain, with x, also the trajectory of x. On the other hand, if a set A is invariant, its indicator function is a constant of motion. Some constants of motion give for each x a set $\{f^{-1}(x)\}$ of phase space measure zero. If this is not the case there is at least one value $f(x)$ such that the invariant set it defines, has positive measure. If it is less than one, the system cannot be ergodic, as the motions within the invariant sets are confined to it. Now, on the other hand, let us assume that our system is not ergodic. There is, according to definition, at least one invariant set A having measure strictly between zero and one. Its indicator function is a constant of motion. Furthermore, it is a function which cannot be determined from the total energy of the system.

Let us next consider time averages. They are phase functions $\hat{f}(x)$. We first took finite time averages from t_1 to t_2, subsequently let $t_1 = 0$ and $t_2 \to \infty$. Now, let us note that $\hat{f}(x) = \hat{f}(T_t x)$. Let t' be a fixed moment of time. Then $\hat{f}(x) = \lim_{t\to\infty} 1/t \int_0^t f(T_t x)\,dt = \lim_{t\to\infty} 1/t[\int_0^{t'} f(T_t x)\,dt + \int_{t'}^t f(T_t x)\,dt]$. The left summand multiplied by $1/t$ goes to 0 as $t \to \infty$, so that, noting that the constant t' can be deducted from t without altering the limit as $t \to \infty$, we have $\hat{f}(x) = \lim_{t\to\infty} 1/(t - t') \int_{t'}^t f(T_t x)\,dt = \hat{f}(T_{t'} x)$. Therefore time averages of stationary systems are constants of motion. If the system is ergodic, time averages have constant values. Time averages of indicators coincide with probabilities, so that probabilities are constants of motion.

Above we saw that failure of ergodicity amounts to the existence of more than one constant of motion. In the other direction, constants of motion can be turned into collections of invariant sets. These again lead to failure of ergodicity if they have measure strictly between zero and one. Let f be a strictly increasing real function of x and a constant of motion. Even if all of $\{f^{-1}(x)\}$ have measure zero, it is easy to define a new invariant $f_a(x)$ such that the sets $A_1 = \{x|f(x) \leqslant a\}$ and $A_2 = \{x|f(x) > a\}$ both have positive measure. The system cannot be ergodic. In conclusion, ergodicity is (at least almost) equivalent to having only one constant of motion for the system ("almost" because there might be additional constants of motion from which no invariant sets of measure strictly between 0 and 1 can be constructed).[17]

For classical systems having a phase space of dimension n, each set of values of k independent constants gives a subspace of dimension $n - k$. States x for which the constants have these values, are constrained to remain in the said subspace. Each additional independent constant reduces the dimension of the allowed subspace by one. Let us call a set of independent constants of motion $f_1, \ldots, f_k$ *complete* if all other constants of motion can be expressed in terms of the f_i. By this is meant that if f is a constant, there is a function g such that $f(x) = g(f_1(x), \ldots, f_k(x))$. To be able to determine $f(x)$, it is enough to have the values of the $f_i(x)$. Specifically, any time average can be determined from the f_i. Let us next assume that our system is ergodic, so that it has only one independent constant of motion, the total energy $H(x)$. On the other hand, in this case time averages also are constants of motion. If $\hat{f}(x)$ is a time average, there is a function g such that $\hat{f}(x) = g(H(x))$. Since $H(x)$ is a constant over phase space, $\hat{f}(x)$ too must be.[18] Probabilities of ergodic classical systems are multiples of total energy. In the other direction, if time averages fail to be constant over phase space, one can use them for grouping trajectories into invariant classes having measure strictly between zero and one.

The uniform probability singled out by the total energy of an ergodic system is called *microcanonical*. It can be seen as a limiting case of two opposing situations. First, no system is really isolated. The total energy changes in value if the system interacts with its surroundings. By suitable assumptions on the latter (heat bath) it is possible to derive a distribution for the system belonging to the exponential family. It is known as the canonical distribution. Secondly, it is possible to obtain a generalized microcanonical probability by allowing more than one independent constant of motion.[19] Obviously, such a generalization relates directly to the ergodic decomposition discussed above.[20] Some remarks will be made on this in the next section.

The reader might get the impression that the present approach explains only uniform probability distributions from a physical criterion. This is not the case. Imagine a point in straight motion (with geometric collision at borders) in an area of the form of a stadium. It is known to be ergodic, so that the point should be seen approximately equally often on the right side as on the left side of the stadium (symmetrically divided). One asks, why could the point not 'slow down' on the left side, so as to produce different time averages for the two parts of equal areas? The answer to this objection[21] shows the intrinsic character of the phase space measure. Uniformity holds for Hamiltonian position and momentum coordinate space. 'Slowing down on the left side' of course decreases momentum (mass times velocity) so that the phase space for the 'left part' is smaller than the right, exactly in proportion to the respective time averages. In configuration (position coordinate) space, the distribution is not uniform. This kind of physical argument for a non-uniform probability was first given by Hopf in the mid-thirties. His examples were a bit more realistic than the one above. For example, a roulette wheel with a friction varying in direction. The problem first has to be transformed into the right coordinate space; a uniform distribution is derived, and the distribution of the original variables recovered from this.[22]

Uniform distributions fail (remembering that they may not always *appear* uniform as shown above) either because the world changes (the system is not isolated well enough), or because it accepts of more than one independent invariant of motion. The latter situation will now be used to clarify certain basic notions.

5. CAUSAL FACTORS, PROBABILISTIC CAUSALITY AND EXPLANATION

Let us suppose that we have a system whose state space is a subset of R^n. The system has a degree of freedom n. It is supposed to be isolated so that total energy is a constant of motion. This reduces the space of possible states to a surface of dimension $n - 1$. Any further independent constants of motion reduce the dimension further, until a 1-dimensional subspace (a single trajectory of the system) is reached. Determination of the values of all the constants will therefore lead to a unique solution of the system's equation of motion. Earlier, we identified all physical properties as constants of motion. In the above case, the system's microscopic state is determined from the values of n observable physical properties. The system is deterministic. But if it is impossible to identify the microscopic state through macroscopic observations, for the reason that the number of independent constants of motion is less than $n - 1$, the system is *not* deterministic in a physical sense.[23]

It follows from the above view that causality is no more identified, in classical systems, with the uniqueness of solutions for given initial conditions of particular differential equations. Rather, only those properties appear in causal relations with which a distinction can be made to other systems of identical structure. Now, it is clear that if a system consists of a large number of parts (e.g., molecules), it has a high degree of freedom, but its macroscopic properties cannot usually be equally numerous. This is expressed by saying that the number of independent constants of motion k is much less than the degree of freedom n.

Each specification of values of a complete set of invariants of motion $f_1, \ldots, f_k$ gives an ergodic component. The state space is thus decomposed into parts which cannot be further decomposed by any macroscopic criterion. A family of uniform distributions parametrized by the values of $f_1, \ldots, f_k$, the generalized microcanonical distribution, gives the statistical laws for each of these parts.

In the special case of $k = n - 1$, it is possible to control or observe the values of the f_i so as to give a unique trajectory. In the other cases, it is only possible to determine which of the different possible statistical laws govern the occurrence of events. In either case, we suggest that *causal factors* in an experiment are identified as constants of motion. In the deterministic case, control of causal factors makes a particular trajectory occur. In the statistical case, these factors make events to occur with a particular statistical law. This was identify as *probabilistic causality*.[24]

The statistical and deterministic forms of causality appear as different special cases of the same theoretical scheme. In the usual (deductive-nomological) form of scientific explanation, as applied in broad terms to the above, we have in the deterministic case as a general law a dynamical law of motion. As particular facts there appear initial conditions. As the complete set of invariants numbers the degree of freedom minus one, a unique solution is determined. Given the general law and the values of the invariants, the dynamical law relates to each other particular states of a single trajectory. If the occurrence of some future event of interest is determined by the microstate at that time, its explanation consists of a law, values of invariants, and a (contingent) observation of the state at some earlier time. In the statistical case, initial conditions are again given as values of a complete set of invariants. The dynamical law is replaced by a family of conditional distributions, the generalized microcanonical distribution. The situation is not an alternative to the deterministic case. The latter can be recovered from the former by constructing a family of measures, each concentrated on a trajectory determined by the values of invariants. The dynamical law will transform each of these sets of probability one into itself while preserving the probability one. In the general case the role of dynamics is the same: to single out a family of probabilities that are preserved under the motion. The statistical and deterministic cases both fall under a scheme with, first, a parametric family of probability laws, second, a fixing of values of the parameters, and third, a particular probability law that follows from the first together with the second point. No difference of principle is seen between explanation in the deterministic and statistical cases.

It is of interest to note how some of the earlier literature on causal factors in experiment and on probabilistic causality and explanation relates to our suggestions. The point of departure is usually tied to the notion of indecomposability. We saw that once an ergodic component has been reached, no further specification of the conditions of an experiment will be able to make a difference to the statistical laws. In earlier literature, the latter circumstance was taken to mean that a form of indeterminism, of the kind best exemplified by quantum theory, has to obtain here. Reichenbach, and, following him, Salmon, are afraid that the proper homogeneous reference class goes down to a point. Fisher retreats to subjectivism which makes 'stratification' relative. Hempel thinks 'maximal specificity' makes probabilities disappear unless epistemic relativity or indeterminism is accepted. For Niiniluoto, probabilities are insensitive to informational increase only under indeterminism, and so on.[25] The best way to cut short the above complex of claims is to adopt the notion of determinism suggested above: that the number of independent constants of motion matches the degree of freedom. This makes some classical systems indeterministic, but saves all of the conceptual analyses of the above writers and many others, except for this particular point about classical systems. The reader will have little difficulty in relating to our scheme their various suggestions concerning causal factors, probabilistic causality, and probabilistic explanation.

Two final points. Above in Section 3, it was suggested that if an event affects the experimental setting of another event, causal independence is violated. Events can be probabilistically dependent even if they are causally independent in the above sense. A familiar example, entirely within our scheme, is obtained from exchangeability. In the usual case of tossing a coin with unknown bias, successive events are positively correlated under exchangeability. However, the result of one toss obviously has no causal effect whatsoever on the subsequent tosses.[26] Instead, the events are causally dependent on the (unknown) parameter characterizing the bias. Conditional on its value, the events are independent as is well known.

Secondly, we suggested that probabilistic explanation is of the following form: There is a family of parametric probabilistic laws which is derived from a structural description of the experimental arrangement.[27] Values of causal factors are fixed. It follows that a particular probabilistic law obtains. If we read it in terms of time averages, we have that with probability one, the limit of time average (or relative frequency) equals a certain number which is the probability of the event in question. If we consider probability one a limiting form of high probability, we have that the finite time average (finite relative frequency) is with high probability near the probability of the event. Therefore, we may consider that a further 'inductive inference' has been drawn from the particular probabilistic law.

Notions of causal factors, probabilistic causality and explanation have now been established. The physical background of these notions ensures their scientific applicability. Their use is not restricted by this background. They form part of a general theory of repetitive experimentation which is seen emerging from the abstract form of ergodic theory.

University of Helsinki

NOTES

[1] See, for example, Tolman (1938).

[2] A discussion of the historical origins and further development of this notion can be found in our (1987).

[3] One of the original contributors, Gibbs himself, suggested that one should adopt the time average view if the ensemble appears lacking in physical meaning.

[4] Two of the first have been presented earlier, in our (1982) and (1983b) whereas the two latter are new.

[5] See Birkhoff and von Neumann (1936). This function should in fact be identified with one having uniformly value 1.

[6] See for example Farquhar (1964) or Khintchine (1949) for proofs. Cornfeld, Fomin and Sinai (1982) is a recent extensive presentation of ergodic theory.

[7] See Grad (1967), p. 67–68 for these interpretations.

[8] This is the point of view of Lewis (1960) and, following him, Truesdell (1961).
[9] A related argument is given in Grad (1967), p. 72.
[10] See Halmos (1939), p. 465.
[11] He called this the principle of exclusion of gambling systems. Other frequentist thinkers have come to similar ideas. Fisher, for example, thinks of probabilities as relative frequencies in experiments. These should be insensitive to "recognizable stratification" of the basic population. He thought that this would be a subjective criterion (Fisher 1973, p. 35). It suffices to note the objective character of ergodicity as a "stratification rejector".
[12] These can be found in Billingsley (1965) or Ornstein (1974).
[13] There is an interesting similarity to the results of Per Martin-Löf (1966) on the theory of random sequences. It is a modern version of von Mises' theory. There it can be shown that a certain condition of randomness (maximum complexity) also gives unique limits of relative frequencies, so that there von Mises' second postulate for frequentist probability is redundant. Moreover, in Martin-Löf's theory, the uniqueness holds uniformly. We have here two cases, ergodicity & co and maximum complexity, both leading to the fulfilment of von Mises' second requirement. No account of the relations of these two approaches exist. Specifically, it would be interesting to see how the complexity theory applies to classical dynamical systems. Some indications in this direction can be found in the book of Lichtenberg and Lieberman (1982).
[14] This view goes back to at leat Grad (1952), p. 460, who says that stationarity "can be verified *a priori*" when successive observations of the same system are made.
[15] Some of the reasons can be gathered from de Finetti (1970), chapter 9, p. 212.
[16] According to standard terminology, for example, Feller (1971), p. 238, a law of large numbers concerns independent, identically distributed random variables. In Doob (1953), p. 464, the ergodic theorem is introduced as a 'law of large numbers for strictly stationary processes'. A concise proof of the probabilistic ergodic theorem can be found in Doob's book.
[17] A precise statement can be found in Lewis (1960).
[18] This is, of course, only repeating an earlier conclusion about the constancy of $\hat{f}(x)$ over X in ergodic systems, but the argument used here generalizes to non-ergodic cases.
[19] This has been first done by Grad (1952), and by Lewis (1960) in a general measure theoretic formulation.
[20] It seems that the situation has not been treated in these terms before. The mathematical details of the ensuing discussion can be gathered from Lewis (1960).
[21] Raised by Prof. Patrick Suppes in discussion.
[22] See our (1983a) for a discussion and references.
[23] Note that the traditional view of determinism in classical systems is different. It is not connected to physically meaningful properties, but to a certain idealistic view of the role of mathematical notions.
[24] Compare Grad (1952), p. 460 and (1967), p. 70 for the physical context of these ideas. The exact construction of the parametric family of microcanonical distributions can be found in Lewis (1960).
[25] See Niiniluoto (1982) for review of the above topic.
[26] See Feller (1968), pp. 121–124 for a discussion of this point.
[27] Talk about arrangement, trial, etc. could as well be replaced by a less operational terminology of system, event, and so on.

REFERENCES

Billingsley, P. (1965) *Ergodic Theory and Information* (New York: Wiley).

Birkhoff, G. and J. von Neumann (1936) 'The logic of quantum mechanics', *Annals of Mathematics* **37**, pp. 823–843.

Cornfeld, I. P., S. V. Fomin, and Ya. G. Sinai (1982) *Ergodic Theory* (Berlin: Springer).

Doob, J. L. (1953) *Stochastic Processes* (New York: Wiley).

Farquhar, I. E. (1964) *Ergodic Theory in Statistical Mechanics* (New York: Wiley).

Feller, W. (1968, 1971) *An Introduction to Probability Theory and Its Applications*, vol. 1, 3rd ed. (1968) and vol. 2, 2nd ed. (1971) (New York: Wiley).

de Finetti, B. (1970) *Probability, Induction and Statistics* (New York: Wiley).

Fisher, R. A. (1973) *Statistical Methods and Scientific Inference*, 3rd ed. (New York: Hafner Press).

Grad, H. (1952) 'Statistical mechanics, thermodynamics, and fluid dynamics of systems with an arbitrary number of integrals', *Communications in Pure and Applied Mathematics* **5**, pp. 455–494.

Grad, H. (1967) 'Levels of description in statistical mechanics and thermodynamics', in Bunge, M. (ed.), *Delaware Seminar in the Foundations of Physics* (Berlin: Springer).

Halmos, P. R. (1939) 'Invariants of certain stochastic transformations: The mathematical theory of gambling systems', *Duke Mathematical Journal* **5**, pp. 461–478.

Khintchine, A. I. (1949) *Mathematical Foundations of Statistical Mechanics* (New York: Dover).

Kingman, J. F. C. (1978) 'Uses of exchangeability', *The Annals of Probability* **6**, pp. 183–197.

Lewis, R. M. (1960) 'Measure-theoretic foundations of statistical mechanics', *Archive for Rational Mechanics and Analysis* **5**, pp. 355–381.

Lichtenberg, R. and R. Lieberman (1982) *Regular and Stochastic Motion* (Berlin: Springer).

Martin-Löf, P. (1966) 'The definition of random sequences', *Information and Control* **9**, pp. 602–619.

Niiniluoto, I. (1982) 'Statistical explanation', in G. Floistad (ed.), *Contemporary Philosophy. A New Survey*, vol. 2, pp. 157–187 (The Hague: Martinus Nijhoff).

Ornstein, D. S. (1974) *Ergodic Theory, Randomness and Dynamical Systems* (New Haven: Yale University Press).

von Plato, J. (1982) 'The generalization of de Finetti's representation theorem to stationary probabilities', in P. Asquith and T. Nickles (eds.), *PSA 1982*, vol. 1, pp. 137–144 (East Lansing, Michigan: Philosophy of Science Association).

von Plato, J. (1983a) 'The method of arbitrary functions', *The British Journal for the Philosophy of Science* **34**, pp. 37–47.

von Plato, J. (1983b) 'The significance of the ergodic decomposition of stationary measures for the interpretation of probability', *Synthese* **53**, pp. 419–432.

von Plato, J. (1987) 'Probabilistic physics the classical way', in Gigerenzer, G., L. Krüger, and M. Morgan (eds.), *The Probabilistic Revolution*, vol. 2 (Cambridge, Massachusetts: MIT Press).

Tolman, R. C. (1938) *The Principles of Statistical Mechanics* (Oxford).

Truesdell, C. (1961) 'Ergodic theory in classical statistical mechanics', in P. Caldirola (ed.), *Ergodic Theories*, pp. 21–56 (New York: Academic Press).

INDEX OF NAMES

INDEX OF SUBJECTS